AF357828

Selected Papers from XXI SPB National Congress of Biochemistry 2021

Guest Editors

Manuel Aureliano
M. Leonor Cancela
Célia M. Antunes
Ana Cristina Rodrigues Costa

Basel • Beijing • Wuhan • Barcelona • Belgrade • Novi Sad • Cluj • Manchester

Guest Editors

Manuel Aureliano
FCT
University of Algarve
Faro
Portugal

M. Leonor Cancela
FMCB
University of Algarve
Faro
Portugal

Célia M. Antunes
DCMS, ESDH
University of Évora
Évora
Portugal

Ana Cristina Rodrigues Costa
DCMS, ESDH
University of Évora
Évora
Portugal

Editorial Office
MDPI AG
Grosspeteranlage 5
4052 Basel, Switzerland

This is a reprint of the Special Issue, published open access by the journal *BioChem* (ISSN 2673-6411), freely accessible at: https://www.mdpi.com/journal/biochem/special_issues/Selected_Papers_from_XXI_SPB_National_Congress_Biochemistry_2021.

For citation purposes, cite each article independently as indicated on the article page online and as indicated below:

Lastname, A.A.; Lastname, B.B. Article Title. *Journal Name* **Year**, *Volume Number*, Page Range.

ISBN 978-3-7258-3952-0 (Hbk)
ISBN 978-3-7258-3951-3 (PDF)
https://doi.org/10.3390/books978-3-7258-3951-3

Contents

Selected Papers from XXI SPB National Congress of Biochemistry 2021

About the Editors

Manuel Aureliano

Manuel Aureliano is a Full Professor of Biochemistry (Aggregation in Inorganic Biochemistry) with the Faculty of Sciences and Technology of the University of Algarve, Faro, Portugal, teaching biochemistry, inorganic biochemistry, muscle contraction and advanced chemistry and biochemistry. He undertook his undergraduate, master, and PhD studies at the University of Coimbra, Coimbra, Portugal. He was the Director of the Biochemistry Degree at the University of Algarve for about two decades, from 1998 to 2013 and from 2021 to 2025. Currently, he is at the Algarve Centre for Marine Sciences (CCMAR-Algarve), and his recent projects include the study of applications of polyoxometalates in the environment and medicine. Aureliano was selected as an "Outstanding Reviewer" (top 10 reviewers) for the journal *Metallomics* over 3 consecutive years (2017, 2018, and 2019). In 2021, 2022, 2023, and 2024, he was included in the "Worlds Top 2% Scientists list" (impact-career and year). Recently (2024), he was the winner of the 3rd edition of the "UAlg Researcher Award" awarded by the University of Algarve.

M. Leonor Cancela

Leonor Cancela (LC) is a Full Professor with the Faculty of Medicine and Biomedical Sciences of the University of Algarve/UALG, a member of the Algarve Biomedical Centre, and a founding member of the Centre for Marine Sciences/CCMAR. She undertook her undergraduate/graduate studies at University Paris VI- Pierre et Marie Curie and PhD work at INSERM/Hospital Lariboisière in Paris. She was hired as a Postdoctoral Researcher by the Univ. of California between Jan 1986 and Jan 1988 at UC Riverside and Feb1988 and Nov1992 at UC San Diego. She obtained a position at UALG in Dec 1992 and became a Full Professor in 2006. LC was elected President of the Portuguese Biochemical Society between 2005 and 2008, has been a member of the board of the Portuguese Society of Genetics since 2010, and was elected as President in January 2020 to present. She has served as the Director of the PhD (since 2008) and Bachelor (2006–2024) programs in Biomedical Sciences, Director of the UALG Doctoral College (2020 to the present), and President of the Scientific Council of the Department, now the Faculty of Medicine and Biomedical Sciences (2019–present). She implemented the BIOSKEL research laboratory that she has led since 1994; has coordinated over 150 undergraduate, PhD, and postdoctoral fellows; and published over 200 articles in international refereed journals, 15 book chapters, and 7 patents. She is a co-founder of Technozeb, a Business Unit of Technophage SA involved in the discovery of molecules with therapeutic interest using zebrafish. LC is regularly invited to give seminars and integrate juris and evaluation panels in national/international institutions and agencies, including as the chair of ERC committees and sessions at national and international congresses. Her research interests focus on molecular, environmental, and epigenetic determinants of skeletogenesis in health and disease and the use of zebrafish as a model for human pathologies.

Célia M. Antunes

Célia M. Antunes (CMA) is an Associate Professor with the Department of Medical and Health Sciences, the School of Health and Human Development (ESDH), University of Évora (UÉvora), Portugal. She is the Vice-Director of ESDH (2021-present), a member of the Council of the Centro Académico Clínico do Alentejo (C-TRAIL) since 2022, and a founding member of the Center of Sci-Tech Research for the Earth System and Energy (CREATE, UÉvora).

Her teaching experience covers biochemistry, immunology, immunity, and the environment and technology of tissue and cell culture. At UÉvora, she has been the Director of the BSc in Biomedical and Health Sciences (2022–present) and Biochemistry degree (BSc 2005–2016; Master's 2018–2022) and a member of the Course Committee of BSC in Human Biology (2009–2021) and PhD in Biochemistry (2009–present). She implemented and has been coordinating the Cell Culture Lab (ESDH) and the Bioaerols and Health and Environment Research Labs (CREATE). She has been a superviser or co-supervisor for over 100 undergraduate, master, PhD, and post-doc fellows. She is the author or co-author of over 50 peer-reviewed papers, 1 patent, 4 book chapters, and several reports and has edited 1 book. She led the creation of the PolenAlert platform. CMA integrated the Board of the Aerobiology and Pollution Working Group of European Academy for Allergy and Clinical Immunology (EAACI) (2012–2014 and 2017–2021), was the Secretary from 2022 to 2024, has been the Chair since June 2024, and has been the Vice-President of the Portuguese Society of Biochemistry since 2022. CMA coordinates with the LifeSpan Chair (since 2024) at UÉvora. Her research interests focus on pathophysiology and environmental public health, particularly environmental health determinants and signalling pathways and biomarkers in allergies.

Ana Cristina Rodrigues Costa

Ana Rodrigues Costa (ARC) is an Assistant Professor at the School of Health and Human Development (ESDH), University of Évora (UÉvora), Portugal, teaching Biochemistry and Toxicology. She has been the Director of the Department of Medical and Health Sciences since 2023. She was a member of the Course Committee of Biochemistry BSc between 2013 and 2021 (and Director between 2015 and 2017) and of the Master's in Biochemistry (2017–present).

ARC graduated in Biochemistry from the Faculty of Sciences of the University of Lisbon and completed her PhD in Biochemistry at the University of Évora. ARC is a Researcher at the Center of Sci-Tech Research for the Earth System and Energy (CREATE, UÉvora), developing research in biochemistry with emphasis on health-related problems, with a focus on biochemical/toxicological approaches to environment–health connections, collaborating with several research groups, particularly contributing to cell-based technologies, including the evaluation of biologic activity and cytotoxicity.

ARC has published 25 articles in scientific journals and 2 book sections and has a registered patent. She has organized several scientific events and supervised one PhD and several MSc and BSC projects, with participation in national and international projects.

Preface

The XXI SPB National Congress of Biochemistry 2021 was held at the University of Évora in Portugal on 14–16 October 2021. Under challenging conditions, due to the COVID-19 pandemic, we managed to organize the National Congress of Biochemistry in a hybrid format, where at least 2/3 of participants came to Évora in person. With the pandemic under control, we carried out the Congress both successfully and safely.

This is the main meeting point for the Portuguese Biochemistry Society (SPB), fostering the discussion and dissemination of high-quality research in biochemistry, both fundamental and applied, taking place in Portugal. The scientific program under the message "Tuning Biochemistry with Life Sciences and Society" covers a broad range of boundaries from molecular mechanisms of diseases to drug discovery, as well as innovative biochemistry projects. Science and innovation was promoted through dialogue, sharing, and healthy confraternization.

Under the scope of the Special Issue on the XXI SPB National Congress of Biochemistry, this e-book shares six reviews, three papers, and one communication. Note that four of these contributions were chosen for the cover of *BioChem* issues. Until now (March 27, 2025), these 10 contributions have garnered a total of 60 citations and 39415 views, indicating an average of 6 citations and 3942 views per publication.

We hope that the present reprint will be used to preserve the memory of future generations of students, researchers, and professors. We also hope that we all find this reprint valuable to our own research work. We would like to thank the Editorial team and reviewers as well as Managing Editor Dr. Nemo Guan for their priceless support and help during the editing process.

Manuel Aureliano, M. Leonor Cancela, Célia M. Antunes, and Ana Cristina Rodrigues Costa
Guest Editors

Review

The Future Is Bright for Polyoxometalates

Manuel Aureliano [1,2]

1   Faculdade de Ciências e Tecnologia (FCT), DCBB, Universidade do Algarve, 8005-139 Faro, Portugal; maalves@ualg.pt
2   Centro de Ciências do Mar (CCMar), Universidade do Algarve, 8005-139 Faro, Portugal

Abstract: Polyoxometalates (POMs) are clusters of units of oxoanions of transition metals, such as Mo, W, V and Nb, that can be formed upon acidification of neutral solutions. Once formed, some POMs have shown to persist in solution, even in the neutral and basic pH range. These inorganic clusters, amenable of a variety of structures, have been studied in environmental, chemical, and industrial fields, having applications in catalysis and macromolecular crystallography, as well as applications in biomedicine, such as cancer, bacterial and viral infections, among others. Herein, we connect recent POMs environmental applications in the decomposition of emergent pollutants with POMs' biomedical activities and effects against cancer, bacteria, and viruses. With recent insights in POMs being pure, organic/inorganic hybrid materials, POM-based ionic liquid crystals and POM-ILs, and their applications in emergent pollutants degradation, including microplastics, are referred. It is perceived that the majority of the POMs studies against cancer, bacteria, and viruses were performed in the last ten years. POMs' biological effects include apoptosis, cell cycle arrest, interference with the ions transport system, inhibition of mRNA synthesis, cell morphology changes, formation of reaction oxygen species, inhibition of virus binding to the host cell, and interaction with virus protein cages, among others. We additionally refer to POMs' interactions with various proteins, including P-type ATPases, aquoporins, cinases, phosphatases, among others. Finally, POMs' stability and speciation at physiological conditions are addressed.

Keywords: polyoxometalates; decavanadate; emergent pollutants; cancer; bacterial resistance; virus infection

Citation: Aureliano, M. The Future Is Bright for Polyoxometalates. *BioChem* **2022**, *2*, 8–26. https://doi.org/10.3390/biochem2010002

Academic Editor: Buyong Ma

Received: 1 December 2021
Accepted: 29 December 2021
Published: 6 January 2022

1. Introduction

Polyoxometalates (POMs) represent a broad class of anionic inorganic clusters of oxoanions of transition metals, such as Mo, W, V, and Nb, with versatile structures resulting in a variety of chemical and physical properties (Figure 1). POM structures can also include other elements, such as P or As, or one of the major metal oxoanions missing and/or substituted by other metals, such as Co or Mn. These inorganic clusters have applications in environmental, chemical, and industrial fields, and are well-known, particularly in catalysis, prevention of corrosion, and macromolecular crystallography, among others [1–6]. The majority of the studies describing POMs' effects in bacteria, viruses, and tumor proliferation, as well as potential drugs for the treatment of several diseases, such as diabetes and Alzheimer's, have been published in the last ten years [1,2,7–18].

Prominent emerging pollutants (EPs) include, for example, antibiotics, antifungals, antidepressants, synthetic hormones, cosmetics, and plastics. Contamination of the environment with those, as well other EP residues, can develop bacteria resistance, itself also an emerging and growing phenomenon worldwide in the 21st century [19–21]. In fact, the resistance of bacteria to antibiotics agents, together with growing cancer incidence all around the world, represent health concerns with increasing relevance. Furthermore, the actual pandemic situation begs for new drugs in the fight against coronavirus, SARS-CoV-2 infection. POMs have been selected [22–24] and followed by an increasing number of researchers as alternative antivirus, antibacterial, and antitumor substances with promising results [1,2,7–18]. In sum, the application of polyoxometalates in the environment and in biomedicine represents two branches of science with rapid growth. Herein, we summarize the reports on the environmental applications of POMs on the eradication of emergent pollutants, and highlight important 21st century studies of POMs' effects and/or targets against cancer, bacteria, and viruses.

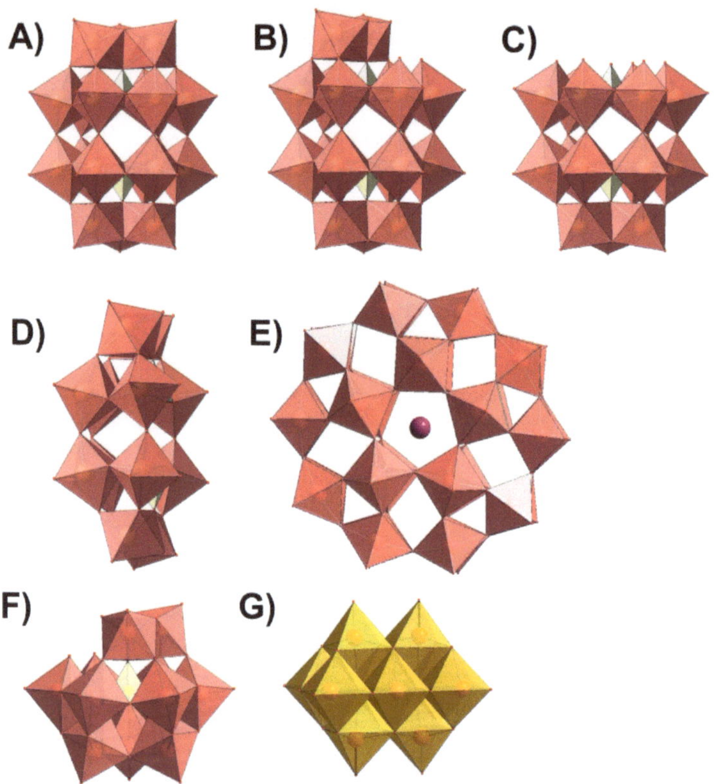

Figure 1. POM-type structures. (**A**) Wells–Dawson (P_2W_{18}); (**B**) mono-lacunary Wells–Dawson (P_2W_{17}); (**C**) tri-lacunary Wells–Dawson (P_2W_{15}); (**D**) hexa-lacunary Wells–Dawson (P_2W_{12}); (**E**) Preyssler (P_5W_{30}); (**F**) mono-lacunary Keggin (MnV_{11}); (**G**) isopolyoxometalate (V_{10}). Color code (**A**–**F**): {WO_6}, reddish; P, yellow; Na, red; (**F**) {VO_6}, reddish; Mn, light yellow; (**G**) {VO_6}, orange.

2. Polyoxometalates against Emerging Health Pollutants

The behavior of humanity has a major impact on the release of organic and/or inorganic pollutants into the environment, and has a profound effect on our lives. In the 21st century, POMs have gained attention as efficient adsorbents and/or green catalysts, and have been used in the development of multifunctional POM materials that could, and can, solve environmental problems, such as water pollution [25–27]. Thus, POMs have been chosen as agents against emergent pollutants. In fact, about 10% (about 1100) of the total of the articles published within the word "polyoxometalate" (POM) (11,000) are studies associated with the environment. In a search on the Web of Science, about 850 articles about POM and degradation can be found, 650 for dyes, 202 for POM and pollutants, 135 for waste, 75 for industrial chemicals, and 70 for wastewater. A lower number of papers were found for POM and pesticides and antibiotics [28–36]. Herein, we describe examples of recent studies about POMs' ability for the degradation of mainly antibiotics, pesticides, and plastics.

Erythromycin and others antibiotics, such as ciprofloxacin, azithromycin, and cefalexin, were also found in effluents and surface waters [19,20]. Ciprofloxacin and erythromycin, together with the macrolide azithromycin, clarithromycin, and the penicillin-type amoxicillin, were included in the surface water watch list under the European Water Framework Directive [20]. More recently, this report was actualized, and the antibacterials sulfamethaxazole and trimethoprim; the anti-fungals clotrimazole, flucozanole, and miconazole; the antidepressant venlafaxine; and the synthetic hormone norethisterone were all added to the 3rd water watch list [21].

POMs were described as a good choice for antibiotic degradation, thus reducing the pharmaceutical environmental impact. In fact, C_3N_4 nanosheet composites loaded with POMs efficiently remove ciprofloxacin, tetracycline, as well as others pollutants, such as bisphenol A [28,29]. Polyoxotungstates (decatungstate, W_{10}) also showed the ability to decompose antibiotics, such as sulfasalazine (SSZ) and one of its human metabolites, sulfapyridine (SPD), with different specificities and rates [30]. W_{10} also has a role in the degradation of pesticides, for example, the ones used for plant growth, namely 2-(1-naphthyl)acetamide (NAD) [31]. A metal-organic frameworks (MOFs) composite of PW12@MFM was shown to display catalytic degradation of sulfamethazine [32]. Besides pharmaceutical drugs, POMs-incorporated frameworks were also found to have applications for the decontamination of dyes, phenolic compounds, and pesticides [33]. Polyoxometalate-based ionic liquid (POM-IL) was used also for the extraction of triazole pesticides, such as hexaconazole, triticonazole, and difenoconazole from aqueous samples [34]. Recent insights in these organic/inorganic hybrid materials, POM-based ionic liquid crystals (POM-ILCs), and their applications (namely on pollutant degradation, including microplastics) have been recently reviewed [35,36].

3. Antibacterial Activity of Polyoxometalates

As referred to above, antibiotic resistance represents a real threat to global public health. The high proportion of bacteria that are resistant to antibiotics is due, at least in part, to antibiotics, as well to other emergent pollutants and environmental contamination. Thus, it is of increasing relevance to pave the way for the exploration of new types of antibiotics for future antibacterial strategies [37]. It was serendipity that caused the first association of POMs with antibacterial activity to be discovered [38]. When comparing the date of the first POMs studies described against viruses and cancer, respectively, at 1971 and 1965 [39,40], the antibacterial activities can be considered recent (1993). However, insights in POMs as anti-microbial agents or adjuvants, as well as POM-ILs as antibacterial coatings, have been attracting interest by offering mechanisms of action different from other recent antibacterial therapies [41–50]. On the other hand, many POMs have poor antibacterial activity, but possess excellent redox activity, and their use as biosensors for bacterial detection has been described [51]. POMs studies reporting the antibacterial activity of POMs are mostly on polyoxotungstates (POTs) and polyoxomolybdates (POMos) (Figure 2). About 80% of these studies were performed in the last 10 years, making them emergent future drugs in the control against pathogenic bacteria [24,52–56]. For example, the POT Preyssler-type $[NaP_5W_{30}O_{110}]^{14-}$ (abbreviated P_5W_{30}) (Figure 1E) showed the highest activity against the Gram-negative human mucosal pathogen *Moraxella catarrhalis*, as well as against *S. aureus* and *E. faecalis* when compared with several others POTs [57]. For *H. pylori*, POMs exhibiting the highest activity were mostly Keggin-type POTs (Figure 1F); polyoxovanadotungstates; and large, highly negatively charged POMs [1].

POTs, such as $[KAs_4W_{40}O_{140}]^{27-}$, (abbreviated As_4W_{40}) and $[P_2W_{18}O_{62}]^{6-}$ (abbreviated P_2W_{18}), are active against H. pylori, as well as metronidazole-resistant H. pylori [58]. Synergetic effects were also found when the antibacterial effect of the β-lactam antibiotic, oxacillin, and the glycopeptide antibiotic, vancomycin, against MRSA and vancomycin-resistant *S. aureus* (VRSA), which is a gram-positive bacteria and one of the most resistant ones around the world, were analyzed in the presence of molybdenum- and tungsten-based POMs, such as the Wells–Dawson type P_2W_{18} (Figure 1A), besides $[SiMo_{12}O_{40}]^{4-}$

(abbreviated SiMo$_{12}$) and [PTi$_2$W$_{10}$O$_{40}$]$^{7-}$ (abbreviated Pti$_2$W$_{10)}$ [59]. Moreover, it was observed that P$_2$W$_{18}$ and Pti$_2$W$_{10}$ were able to change the β-lactam resistance to a β-lactam susceptible profile for the majority of the cases studied [59].

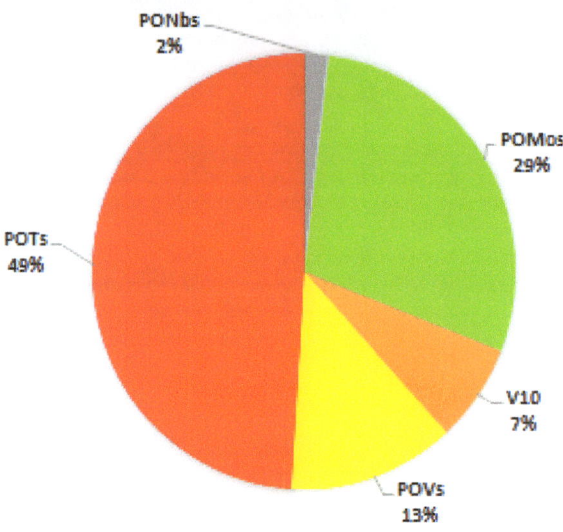

Figure 2. Studies for POMs tested against bacteria. Distribution, in percentage of the different POMs tested for antibacterial activity since 1996. POTs, polyoxotungstates; POMos, polyoxomolybdates; POVs, polyoxovanadates; PONbs, polyoxoniobates.

Besides POTs and POMos, polyoxovanadates (POVs), especially decavanadate (V$_{10}$), were pointed as the most potent against certain bacteria, such as *Streptococcus pneumoniae* [1]. Decavanadate, [V$_{10}$O$_{28}$]$^{6-}$ (abbreviated V$_{10}$), alone has also been described to inhibit the growth of *Mycobacterium smegmatis* and *Mycobacterium tuberculosis* [60]. Moreover, the association of V$_{10}$ with (iso) nicotinamide compounds was described to increase the toxicity towards *Escherichia coli* [61]. Chitosan-V$_{10}$ (CTS-V$_{10}$) complex were also active against *E. coli* and *S. aureus* [52]. Inhibition of *E. coli* growth was also verified for two manganesepolyoxovanadates, namely K$_5$MnV$_{11}$O$_{33}$.10H$_2$O (abbreviated MnV$_{11}$) and K$_7$MnV$_{13}$O$_{38}$_18H$_2$O (abbreviated MnV$_{13}$) [62]. MnV$_{11}$, MnV$_{13}$, and V$_{10}$ were all more potent than vanadate, with 50% maximal growth inhibition concentrations (GI$_{50}$) of 0.21, 0.27, 0.58, and 1.1 mm, respectively, whereas the decaniobate [Nb$_{10}$O$_{28}$]$^{6-}$ (abbreviated Nb$_{10}$) revealed only residual effects on *E. coli* growth [62].

POMs' antibacterial activity, and their comparison with antibiotic drugs are usually measured according to the minimum inhibitory concentration (MIC). Several POMs were tested against *Staphylococcus aureus*, and some showed significant antibacterial activity, exhibiting a MIC < 100 µg/mL, whereas antibiotic drugs have MIC values ranging from 0.001 to 10 µg/mL [1]. Moreover, POMs' MIC relationship with POM size and/or charge of POMs were found for several Gram-positive and Gram-negative bacteria [1]. When the structure-activity analysis was performed for *Streptococcus pneumoniae*, it was demonstrated that this bacterium is especially sensitive to POVs, with decavanadate exhibiting the highest antibacterial activity, whereas for *Heliobacter pylori*, most of the tested POMos were determined to be more active [1]. Table 1 summarizes the MIC values for some POTs, POMos, and POVs being pure, hybrids, and/or composites, which were tested on six bacterial strains, and three Gram-positive bacteria (*Staphylococcus aureus, Bacillus subtilis, Paenibacillus, and three Gram-negative (Escherichia coli, Vibrio, Pseudomona aeruginosa*)) [52,53,56,63–69].

Table 1. Antibacterial activity of POMs, POM-hybrids, and nanocomposites against a series of bacterial strains.

POM/POM-Hybrid	MIC(µg/mL)						Ref.
	S. aureus	*E. coli*	*B. subtilis*	*P. aeruginosa*	*Paenibacillus* sp	*Vibrio* sp	
Polyoxovanadates							
$[V_{10}O_{28}]^{6-}$	50	50	-	-	-	-	[52]
$[V_4O_{12}]^{4-}$	8000						[56]
Polyoxotungstates							
$[PW_{12}O_{40}]^{3-}$	3200						[53]
$[SiW_{12}O_{40}]^{4-}$	3200						
$[BW_{12}O_{40}]^{5-}$	800						[53]
$[PTi_2W_{10}O_{40}]^{7-}$	12,800						[53]
Polyoxomolybdates							[53]
$[P_2Mo_5O_{23}]^{6-}$	6400						[56]
$[MnMo_9O_{32}]^{6-}$	1600						[56]
$[Eu(MoO_4)(H_2O)_{16}(Mo_7O_{24})_4]^{14-}$	400						[56]
Organic-inorganic-POM:							
$[(PhSb^{III})\{Na(H_2O)\}As^{III}_2W_{19}O_{67}(H_2O)]^{11-}$	-	500	125	-	250	125	[64]
$[(PhSb^{III})_2As^{III}_2W_{19}O_{67}(H_2O)]^{10-}$	-	250	62.5	-	125	62.5	[64]
$[(PhSb^{III})_3(B-\alpha-As^{III}W_9O_{33})_2]^{12-}$	-	125	62.5	-	62.5	31.3	[64]
$[(PhSb^{III})_4(A-\alpha-As^VW_9O_{34})_2]^{10-}$	-	62.5	15.6	-	15.6	15.6	[65]
Quinolone-based drug-POM:							
$[Co^{II}(C_{19}FH_{22}N_3O_4)_3][C_{19}FH_{23}N_3O_4][HSiW_{12}O_{40}]$	2.52	2.42	-	-	-	-	[66]
Nanocomposite:							
Bamboo charcoal-POM:							
BC/POM	4	4	4	4	-	-	[67]
Polymer-POM:							
PVA/PEI-POM:							
PVA-PEI-$H_5PV_2Mo_{10}O_{40}$	0.02	2	0.2	0.02	-	-	[68]
Chitosan-POM:							
CTS-$Ca_3V_{10}O_{28}$	12.5	12.5	-	-	-	-	[52]
Polyoxometalate ionic liquids:							
$[N(C_6H_{13})_4]_8[\alpha-SiW_{11}O_{39}]$	10	1000	-	1000	-	-	[69]
$[N(C_7H_{15})_4]_8[\alpha-SiW_{11}O_{39}]$	2	25	-	100	-	-	[69]
$[N(C8H_{17})_4]_8[\alpha-SiW_{11}O_{39}]$	5	50	-	100	-	-	[69]

Mechanisms of Action of Polyoxometalates against Bacteria

The multiple mechanisms of POMs against bacteria and bacterial resistance are not fully understood. The antibacterial effects of POMs and, recently, of POVs, their modes of action, and future perspectives were reviewed [1]. Several possible mechanisms were anticipated, but further studies are required [1,52,56,58,59]. In this section, we resume some of the processes affected by the POMs. Some POVs and POTs are well-known for their interference with the transport system of ions and substrates by inhibiting P-type ATPases, such as Ca^{2+}-ATPase and Na^+/K^+-ATPase [70–73], leading to a disturbance in the molecular ion transport across the membrane, thus affecting bacteria growth [52]. Chitosan-V_{10} were described to induce the inhibition of mRNA synthesis, interfering with protein production, and destroying bacteria metabolism [52]. Suppression of *mecA*-induced mRNA expression, inhibition of the transcription process, and inhibition of the translational process were suggested for P_2W_{18}, $SiMo_{12}$, and PTi_2W_{10} [59]. POMs can also induce the production of reactive oxygen species (ROS), as described for phosphomolybdates and V_{10}-copper(II) tris derivatives, and can further disrupt bacterial cell integrity, showing promising antibacterial activities [74–76].

POMs have also been described to inhibit sialyl and sulfotransferases. Glycosyltransferase catalyzes the transfer of a sialic acid residue to the terminal position of an oligosaccharide chain of glycoproteins and glycolipids [77]. Sulfotransferase catalyzes the transfer reaction of a sulfate group to an acceptor sugar chain on the surface of cells [78]. These modifications of carbohydrate chains play a role in cell–cell recognition, serving as a target for bacterial and also viral infections. In a study with about twenty POMs, three tungstate-based POMs, $[H_2SiNiW_{11}O_{40}]^{6-}$ (abbreviated $SiNiW_{11}O_{40}$), $[Cu_3(PW_9O_{34})_2]^{12-}$ (abbreviated $Cu_3(PW_9O_{34})_2$), and $[SiVW_{11}O_{40}]^{5-}$ (abbreviated $SiVW_{11}O_{40}$), were shown to have a stronger inhibition activity, with IC_{50} values as low as 0.2 nM for the α-2,3-sialyltransferase (ST3Gal-I). Also, three vanadium-based POMs, $[KV_{13}O_{31}(MePO_3)_3]$, $[V_{18}O_{42}(H_2O)]^{12-}$ (abbreviated $V_{18}O_{42}$), and $[V_{18}O_{44}(N_3)]$, have shown to inhibit galactose-3-O-sulfotransferase (Gal3ST-2), with an IC_{50} value of 3 nM for the latter POVs [79].

POMs, such as $SiMo_{12}$ and P_2W_{18}, were suggested to affect the bacterial respiratory system, leading to inhibition of ATP synthesis, thus inducing bacterial death [59]. On the other hand, it was observed that *S. aureus* strains were able to reduce to blue color form POMs, such as P_2W_{18} and $SiMo_{12}$ [59]. Similar blue staining of *E. coli* cells was observed in the presence of vanadium (V) compounds, probably by reduced forms containing oxidovanadium (IV) [62]. Although the molecular mechanisms responsible for the reduction of these POMs by bacteria is not clear, it is reasonable to anticipate that it diverts electrons from cell redox systems, and hinders bacterial growth.

Though it is not clear if POMs can be accumulated by bacteria, there is evidence that some POTs were taken into the bacteria cells [58]. In bacterial cells treated with P_2W_{18}, it was observed that W atoms localized at the periphery of the cells [58]. Changes in cell morphology when *S. pneumoniae* was exposed to POVs [63] were also observed. Similarly, changes at *H. pylori* morphology from bacillary to coccoid upon exposition to several POTs, such as As_4W_{40}, $[KSb_9W_{21}O_{86}]^{18-}$ (abbreviated Sb_9W_{21}), and $[SiVW_{11}O_{40}]^{7-}$ (abbreviated $SiVW_{11}$) were also described [58].

4. Anticancer Activity of Polyoxometalates

As observed above regarding the bacterial studies, the number of studies about POMs with antitumor activities in the past 10 years also represents the majority of the POM anticancer studies: around 87% of all the papers in this field, since the first report by 1965 [40]. Similarly, POT and POMos studies together represent the majority (90%), whereas for POVs and PONbs, lower percentages can be found. In fact, more than 120 articles/papers have been published with POMos and polyoxotungstates POTs in antitumor studies [8,10–13,80–89]. POMs in cancer therapy and diagnostics, their modes of action, and future perspectives were reviewed [2,90–93]. Herein, we summarize examples of

POMs' anticancer effects and putative modes of action, particularly for POVs, for the last five years.

The decavanadate complex proved to exhibit anti-tumor activity against various cancer cell lines, including specific toxicity against human cancer cells, whereas normal human cells were not affected, even for high concentrations of the complex [84]. This complex presented IC_{50} values of 0.72 and 1.8 μM against the human lung adenocarcinoma cell line (A549) and human breast adenocarcinoma cell line (MDA-MB-231), respectively. When V_{10} complex was compared to the antitumor drug cisplatin for its cytotoxicity, it exhibited lower cytotoxicity against A549 than cisplatin, whereas its IC_{50} for MDA-MB-231 cells was 1.7 μM against 700 μM of cisplatin, meaning that it is 400 times more effective. On the other hand, decavanadate alone also showed anticancer activity against HeLa, Hep-2, HepG2, and MDA-MB-231, inducing apoptosis as the process of cell death [84]. The anti-proliferation activity of another POV, V_{18}, was observed to affect the cellular cycle, and to mediate the arrest of MCF-7 cells in the G2/M phase and induction of apoptosis, besides DNA-, BSA-, and HSA-binding [85]. POV studies were also performed with U-87 and human liver SMMC-7721 cancer cells, and cell cycle arrest, DNA damage, and apoptosis were observed [85,94].

Considering all the cancer POMs studies in recent years, only very few were performed in vivo [10,12,87,89,90]. In one of these studies, it was described that the degradability of an organic POMo, based in Mo_6O_{18}, is the key to inhibit human malignant glioma cells (U251), besides having the capacity to cross the blood brain barrier, pointing to a new type of anticancer agent [87]. Another recent study demonstrated the anti-tumor activity of an iron heptatungsten phosphate polyoxometalate complex, $Na_{12}H[Fe(HPW_7O_{28})_2]$ (IHTPO), against large cell lung cancer (NCi-H460), human hepatoma (HepG2), leukemia (K-562), and lung carcinoma (A549) *in vitro*, and against S180 sarcoma transplanted in mice in vivo [10]. Even the cytotoxic effects were only seen at higher concentrations, with IC_{50} values superior to 60 μM, and IHPTO proved to be more efficient against S180 sarcoma transplanted mice. It was concluded that even if this POT exhibited lower antitumor activity than the already approved chemotherapeutic drugs, such as cisplatin, the interesting part is that IHTPO activity might be correlated to an immunomodulatory activity [10].

In another in vivo study, Fu et al. synthesized an amphiphilic organic-inorganic hybrid POT, $[(C_{16}H_{33})_2NCONH(CH_2)_3SiNaP_5W_{29}O_{110}]$ (abbreviated Na-lipidP_5W_{29}), to improve biocompatibility, bioactivity, and biospecificity [12]. Basically, a long chain organoalkoxysilane lipid was grafted into a lacunary Preyssler-type, $[NaP_5W_{29}O_{107}]^{14-}$ (abbreviated P_5W_{29}) in order to produce the desired complex. The hybrid POT, Na-lipidP_5W_{29}, was tested for its antitumor activity against human colorectal cancer cells (HT29), and the results were compared to the parental POT, P_5W_{29}, and to 5-FU. For all concentrations tested, Na-lipidP_5W_{29} exhibited higher inhibitory rates than its parental POT and 5-FU. The cytotoxic effect of the studied POT was also tested against human umbilical vein endothelial cells (HUVECs). Finally, it was suggested that the higher antitumor effect of Na-lipidP_5W_{29} was due to its higher capacity to penetrate the cell, since it can spontaneously assemble into a vesicle [12]. In vivo studies with a Keggin-type POT, $[PW_{11}O_{39}]^{7-}$ (abbreviatedPW_{11}) were also performed against colorectal cancer [87]. To improve bioactivity, and decrease the toxicity effect of this POT, an organometallic derivative of PW_{11} was synthesized and encapsulated to form nanoparticles of Pt^{IV}-PW11-DSPE-PEG2000 (NPs). Results showed that these NPs were more efficient in inhibiting the growth of WT20 cancer cells, and treating human colorectal cancer in mice than cisplatin, pointing once again to a new strategy to fight against cancer [87].

Mechanisms of Action of Polyoxometalates against Cancer Cells

As described above regarding the effects of POMs on cancer cells, several effects were referred, such as cell cycle arrest, apoptosis cell death, and interactions with DNA, among other observations and/or suggestions. However, the multiple mechanisms of action of polyoxometalates as antitumor agents are not yet fully understood. Recently, POMs as anticancer agents were reviewed [2]. In this section, we will resume some of them.

Research is looking for new non-competitive inhibitors of protein kinases, such as the human protein kinase CK2 inhibitors that have already been designated as promising drug targets in cancers [95,96]. POMs, such as P_2Mo_{18}, have been described as non-competitive and potent CK2 inhibitors (IC_{50} = 5 nM); although, due to its instability, it was not possible to know if this POM was responsible for the observed effects [96]. Nevertheless, POMs represent non-classical kinase inhibitors with increasing interest. Recently, aquaporins were also described to be potential protein membrane targets for POTs [97]. Aquaporins (AQPs) were found to be overexpressed in tumors, making their inhibitors of particular interest as anticancer drugs [98]. POTs strongly affect AQP3 activity, and induce inhibition of melanoma cancer cell migration and growth, unveiling their potential as anticancer drugs against tumors, opening a new window in this field of research [97]. P-type ATPases play a crucial role in cellular ion homeostasis, and have been described as potential molecular targets for several types of compounds used in the treatment of ulcers, cancer, heart ischemic failure, among other diseases. Among these compounds, several POMs have been described as PMCA (plasmatic membrane calcium ATPase) and SERCA (sarco(endo)plasmatic membrane calcium ATPase) inhibitors, and the effects compare with other inorganic compounds, as well as with therapeutic drugs [70–73,99,100].

Decavanadate species, and POMs in general, were described as strong inhibitors of phosphatases, such as alkaline phosphatase (ALP) [14,81]. Seven POTs were assessed for their inhibitory effect on alkaline phosphatases ALP, and as putative antitumor agents [14,81]. Abnormal levels of ALP in the serum are detected in cancer patients, since tumors are an abnormal cellular growth proliferating faster than normal cells, and thus, the inhibition of ALP will affect tumor cell metabolism and function. Three different POMs, P_5W_{30}, V_{10}, and the Anderson-Evans type $[TeW_6O_{24}]^{6-}$ (abbreviated TeW_6), with chitosan-encapsulated nanoassemblies were tested as anticancer agents on HeLa cells [15]. The maximum cytotoxicity against HeLa cells was observed for the compound chitosan-P_5W_{30}, which also has higher phosphatase inhibition. It was suggested, in both studies, that the POT with the largest number of tungsten and phosphorus atoms may provide the optimal interaction with the phosphatase [15,81]. Finally, disturbance of antioxidant systems is a plausible anticancer strategy once tumor cells have rapid growth and metastasis. It was found that PW_9Cu concentrations that induced osteosarcoma cells death in vitro also increased ROS, and decreased the reduced gluthatione/oxidized glutathione (GSH/GSSG) ratio in the cells [82]. Moreover, the cytotoxicity of the compound was prevented with the addition of GSH, suggesting that oxidative stress is a mechanism of POMs to induce cancer cell death [82].

5. Antiviral Activity of Polyoxometalates

The number of studies using POMs that address viral infection is comparatively lower than the ones found for cancer and bacteria. Nevertheless, and due to the SARS-CoV2 pandemic [101], the number of studies testing metallodrugs which include POMs for treatment of viral infection has increased in recent years [102–117]. Still, the studies performed so far in the past 10 years represent almost 50% of the total. Among the studies described, and since 1971 [39], the ones using POTs represent the major contribution in this field (75%). Herein, we summarized examples of POMs' antiviral effects and putative modes of action in recent years, highlighting, as above, the in vivo studies.

Considering all the POMs studies published on different types of viruses, it can be observed that influenza, HIV, herpes, and corona are the viruses most studied (Figure 3). Thus, the antiviral activity of POMs has been prevalent in respiratory tract viruses, mainly influenza viruses (Figure 3). A study with one Keggin-type POM, $[SiVW_{11}O_{40}]^{5-}$ (abbreviated $SiVW_{11}$), and two double Keggin-type POMs, $(K_{10}Na[(VO)_3(SbW_9O_{33})_2])$ and $(K_{11}H[(VO)_3(SbW_9O_{33})_2])$, showed activity against dengue virus (DFV), influenza virus

(FluV A), respiratory syncytial virus (RSV), parainfluenza virus (PfluV 2), distemper virus (CDV), and human immunodeficiency virus (HIV) [106]. It was further demonstrated that ($K_{10}Na[(VO)_3(SbW_9O_{33})_2]$) strongly inhibits the binding of the viral gp120 antibodies [106]. P_2W_{18} was also studied on influenza virus (FluV) in MDCK cell line [105]. It was suggested that P_2W_{18} could inhibit the role hemagglutinin A (HA), responsible for the first stage of viral attachment [105]. Thus, the Wells–Dawson-type POM P_2W_{18} is likely to have a dual mechanism of action in the inhibition of FluV replication: it reduces the binding of HA to the host cell membrane glycoprotein receptors, and impedes the fusion of viral particles into the cell [105].

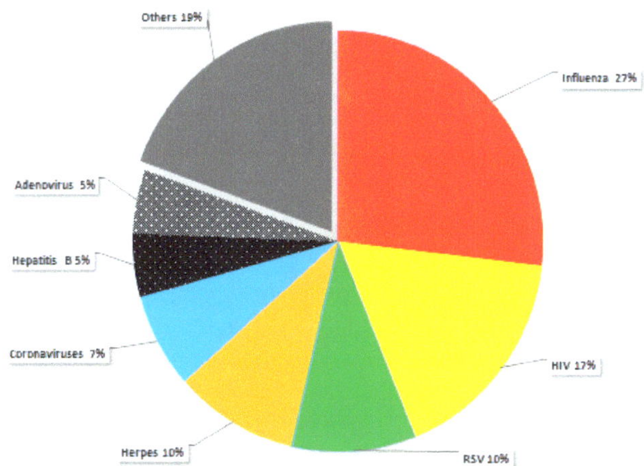

Figure 3. Studies published with all types of POMs on different types of viruses (influenza, HIV, corona, herpes).

It is known that HIV specially targets CD4 molecules present in T lymphocytes, monocytes, and macrophage lineage. It is also well-known that a glycoprotein, denominated gp120, allows it's binding on CD4 cells, and, consequently, the injection of viral material into the host cell [112]. The activity against the human immunodeficiency virus (HIV) was demonstrated for some POTs [102,104,113]. It was suggested that POMs exhibited their antiviral effect by inhibiting the binding of virus to the host cell and/or its penetration [58,59,102,104]. For example, the single Wells–Dawson structure of the compound (α_2-[NMe_3H]_7[CH_3C_5H_4TiP_2W_{17}O_{61}]$), and the double Wells–Dawson of the structure compound ($Na_{16}[Mn_4(H_2O)_2(P_2W_{15}O_{56})_2]$) both inhibited the binding of HIV particles to CD4 cells by blocking the binding of gp120 to SUP-T1 cells [104]. Other studies reported that POMs could inhibit proteases in a non-competitive manner at low micro molar concentrations [14,102,104], thus affecting virus infection.

As referred before for the cancer studies, in vivo POMs antiviral studies remain scarce, and very few studies [114,115] have been performed (Table 2). In this table, we compared the effects of two POTs and two clinically approved drugs in a mouse, the animal model. The Keggin-type POM$[PW_{10}Ti_2O_{40}]^{7-}$ (abbreviated $PW_{10}Ti_2$) shows a survival rate (SR), indicating the percentage of mice that were still alive on day 14 after infection was 97%, for the treatment of HSV-2 virus infection when 25 mg/kg was administrated [114]. Higher survival rates of 90% were also observed upon a variant of influenza virus (FM1) infection for the POT $Ce_2H_3[BW_9{}^{VI}W_2{}^{V}Mn(H_2O)O_{39}]$ (abbreviated $BW_9{}^{VI}W_2{}^{V}Mn$). However, 10 times of the amount administrated was needed to obtain the same rate of survival upon oral administration (100 mg/kg) in comparison with the intraperitoneal mode (10 mg/kg). Lower rates of SR were observed using well-known clinically approved drugs, such as acyclovir (anti-HSV agent) and ribavirin (broad-spectrum antiviral agent) [114,115]. For acyclovir, a 33% survival rate was observed for a 50 mg/kg administration upon HSV-2 virus infection, whereas for ribavirin, a 70% survival rate was obtained after a 200 mg/kg administration for the FM1 influenza virus infection (Table 2). In sum, when comparing the same mode of oral administration upon the same influenza virus infection for the clinically approved drug, ribavirin, and the new antiviral compound, $BW_9{}^{VI}W_2{}^{V}Mn$, it clear that this POT has a higher SR (90% against 70%) for half of the dose administrated (100 mg/kg against 200 mg/kg).

Table 2. In vivo antiviral activity of POMs.

Polyoxotungstates	Virus	Animal	Survival Rate (SR)	Dose (Mode of Administration)	Ref.
$K_7[PW_{10}Ti_2O_{40}]$	HSV-2	mouse	97%	25.0 mg/kg	[114]
$Ce_2H_3[BW_9{}^{VI}W_2{}^{V}Mn(H_2O)O_{39}]$	FM1	mouse	90%	100 mg/kg (o.a.)	[115]
$Ce_2H_3[BW_9{}^{VI}W_2{}^{V}Mn(H_2O)O_{39}]$	FM1	mouse	90%	10 mg/kg (i.p.)	[115]
Clinically approved drugs					
Acyclovir	HSV-2	mouse	33%	50.0 mg/kg	[114]
Ribavirin	FM1	mouse	70%	200 mg/kg (o.a)	[115]

SR = survival rate, indicating the percentage of mice that were still alive on day 14 after infection; (o.a.) = oral administration; (i.p.) = intraperitoneal administration; FM1 = variant of influenza virus; HSV-2 = herpes simplex virus 2.

Mechanisms of Action of Polyoxometalates against Viral Infection

It was suggested that some POMs could inhibit the replication of HIV [102]. Certain POMs, such as $Cs_2K_4Na[SiW_9Nb_3O_{40}]$ (abbreviated $SiW_9Nb_3O_{40}$), can act directly on hepatitis C virus (HCV) virion particles, and destabilize the integrity of its structure [115]. It was further suggested that POMs could specifically inhibit HCV infection at an early stage of its life cycle [103]. Note that the most cited paper regarding decavanadate (V_{10}) in biology is the interaction of V_{10} in a spatially selective manner within the protein cages of virions [113]. Besides preventing the formation of virions, decavanadate is also able to inhibit viral activities by preventing the virus-cell host binding [113].

As mentioned before, the inhibition of catalytic reactions promoted by sialyltransferases and sulfotransferases would affect the carbohydrate chains in glycoproteins that play a major part in cell–viral recognition, serving as a target for viral infections. Thus, by targeting virus membrane proteins, POMs would affect the early stage of viral infection. Moreover, some POMs are likely to have a dual mechanism of action in the inhibition of FluV replication: interacting with hemagglutinin A (HA), responsible for the first stage of viral attachment (Figure 4), and inhibiting the fusion of viral particles into the cell [110].

HIV is known to specially target CD4 molecules present in T lymphocytes, monocytes, and macrophage lineage. A glycoprotein, denominated gp120, which is located on the surface of the virus, is the principal weapon of HIV because it allows its binding on CD4 cells, and the injection of the viral material into the cell [112]. Other studies reported that POMs could inhibit proteases in a non-competitive manner at low micro molar concentrations [14,102,104]. The HIV-1 protease is important for the maturation of protein components of an infectious HIV virion; thus, its inhibition could be responsible for the anti-HIV effect of POMs. For SARS-CoV, it was referred that POMs interact with the 3CL^{pro} protein, affecting virus proliferation [116,117].

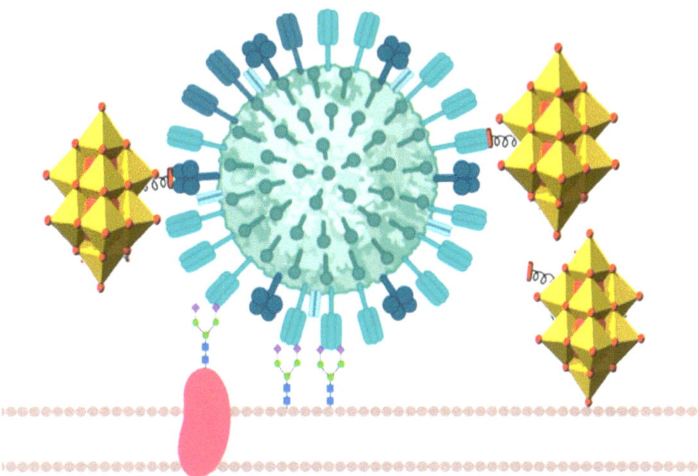

Figure 4. POM putative interactions with viral membrane proteins, such as hemagglutinin, preventing the early stage of infection. Moreover, interaction with neuraminidase would prevent a later stage of infection. Color code: glycoprotein, dark pink; hemagglutinin, dry green; neuraminidase, dark green; POM, yellow.

6. POMs Stability and Speciation in Solution

POMs are clusters of units of oxoanions of transition metals that can be formed upon acidification of neutral solutions of Mo, W, V, and Nb. Once formed, some POMs, such as POVs and PONbs, have shown to persist in solution, even in the neutral and basic pH range [118]. Speciation and/or stability studies are not usually taken into account when studies of the effects of POMs in biological systems are performed [91]. This speciation topic was recently reviewed and highlighted [119]. Nevertheless, it is well-known that some POMs are not stable in many of the experimental conditions, and references to POMs' stability are mentioned, even after it is realized that the POM added to the medium reaction is no longer there, or at least is partially decomposed 5 min, 30 min, or 24 h after incubation [91,118–124]. As mentioned above, in some studies with POTs and bacteria, a change at the color of the medium was detected, suggesting POT decomposition, whereas oxidation reduction reaction could be induced through bacterial metabolism [59,62]. That means that the biological effects observed could not, at least in part, be totally attributed to the POM that was added to the medium.

The isopolyoxovanadate decavanadate (V_{10}) is perhaps the most studied regarding its stability at biological conditions [118–124]. V_{10} demonstrates many important roles in fundamental biological processes [91,122,125–129]. It was suggested that the presence of proteins, such as actin and Ca^{2+}-ATPase, can significantly (5 and 3-fold) improve the stability of V_{10}, whereas no changes were observed for myosin and lipid structures namely liposomes [124]. In fact, it was observed that the half-life time of V_{10} decomposition increases from 5 to 27 h in the presence of G-actin, the monomeric form of actin (Figure 5). Further studies point out specific binding sites for V_{10} at G-actin, in the absence and in the presence of the natural ligand, ATP [130]. Thus, POMs' stability and speciation in the presence of biomolecules will be essential for understanding and deducing their fundamental roles in biology, and their applications in medicine.

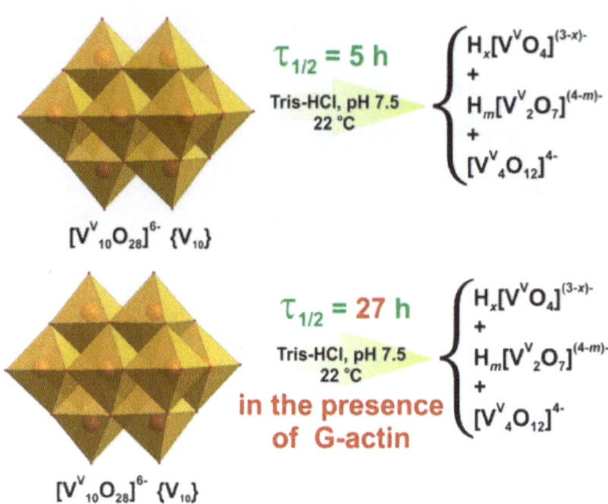

Figure 5. POM species in the presence of macromolecules. Decavanadate, an isopolyoxovanadate, was found to be stabilized upon interaction with G-actin and Ca^{2+}-ATPase [124].

7. Conclusions

POMs are interesting compounds with a diversity of structures. POMs have been studied and applicated in a large variety of fields. POM applications in environmental and biomedical sciences are promising. Herein, we highlighted POMs' degradation of emergent health pollutants, and their anticancer, antibacterial, and antiviral effects, and mechanisms of action.

POMs, as organic/inorganic hybrid materials, seem to be a good choice for emergent health pollutant degradation, including microplastics. In fact, POMs efficiently induce antibiotics degradation, thus reducing the pharmaceutical environmental impact. Emergent pollutant antibiotics, such ciprofloxacin, tetracycline, sulfasalazine, among others, as well as others pollutants, such as bisphenol A, can be decomposed in the environment using POMs.

POMs showed antibacterial activity by inducing morphological changes through cytoskeleton interactions, interfering with the ionic transport systems, leading bacteria to death. Moreover, POMs have demonstrated activity against antibiotic-resistant bacteria. The majority of the studies have addressed the potential of POMs to control bacterial growth, and some explored the mechanism of action of these compounds; however, several aspects of the virulence and community life of bacteria were not yet explored. POMs showed their antiviral activity against influenza, HIV, and several other viruses. POMs are able to inhibit virus infection by preventing the virus-cell host binding at an early stage of the viral infection by targeting viral membrane proteins.

In general, POMs inhibit the growth of tumor cells through apoptosis. Generation of oxidative stress, inhibition of the kinases, ATPases, and/or aquaporin's function, among others, was proposed by some authors. For example, decavanadate and other POMs exhibited anti-tumor activities through the inhibition of ALP, kinases, and P-type ATPases, whereas, so far, no decavanadates were found to inhibit HDAC or ecto-ATPases. POMs' effects on different types of cancer cells were also found to be different.

In sum, several types of POMs have proved to be efficient against viruses, bacteria, and tumor cells, besides their applications in the decomposition of emergent health pollutants. POTs are among those POMs that have been the subject of the largest number of studies on anticancer, antiviral, and antibacterial activities. POM molecular targets are unveiling their potential as anticancer, antibacterial, and antiviral drugs of the future, thus opening new windows for future research in these fields. Further studies involving interdisciplinary teams must be conducted to understand which POM is better-tuned against a particular disease or infection. Thus, the future is bright for polyoxometalates.

Funding: This study received national funds through FCT—Foundation for Science and Technology through project UIDB/04326/2020.

Acknowledgments: Thank to Nadiia I. Gumerova and to Annette Rompel, from Wien University, for the POMs figures structures.

Conflicts of Interest: The author has no conflict of interest to declare.

Abbreviations

A549	Human lung cancer cells
ALP	Alkaline phosphatase
As_4W_{40}	$[KAs_4W_{40}O_{140}]^{27-}$
ATPase	Adenosine triphosphatase
BWCN	$(Himi)_2[Bi_2W_{20}O_{66}(OH)_4CO_2(H_2O)_6Na_4\ (H_2O)_{14}]\cdot17H_2O$
BW9VIW2VMn	$Ce_2H_3[BW_9{}^{VI}W_2{}^VMn(H_2O)O_{39}]$
Ca^{2+}-ATPase	Calcium adenosine triphosphatase
CTS-$Ca_3V_{10}O_{28}$	Chitosan-$Ca_3V_{10}O_{28}(NH_4)_6$
$Cu_3(PW_9O_{34})_2$	$[Cu_3(PW_9O_{34})_2]^{12-}$
gp120	Glycoprotein expressed by HIV
HCB	Hepatitis B virus
HCMV	Human cytomegalovirus
HCV	Hepatitis C virus
HDAC	Histone deacetylase
HeLa	Human cervical cancer cells
Hep-2	Human Larynx carcinoma cell line
HepG2	Human hepato-cellular carcinoma
HIV	Human immunodeficiency virus
HUVEC	Human umbilical vein endothelial cells
IC_{50}	Half inhibitory concentration
K-562	Human myelogenous leukemia
MCF-7	Human breast cancer cells
MDA-MB-231	Human breast adenocarcinoma cell line
MDCK	Madin–Darby canine kidney
mecA	Gene that codes for PBP2′
MnV_{11}	$K_5MnV_{11}O_{33}\cdot10H_2O$
MnV_{13}	$K_7MnV_{13}O_{38}\cdot18H_2O$
Na-lipidP_5W_{29}	$[(C_{16}H_{33})_2NCONH(CH_2)_3SiNaP_5W_{29}O_{110}]$
Nb_{10}	$[Nb_{10}O_{28}]^{6-}$
POMos	Polyoxomolybdates
POMs	Polyoxometalates
PONbs	Polyoxoniobates
POTs	Polyoxotungstates
POVs	Polyoxovanadates
PVA/PEI	poly(vinylalcohol)/polyethylenimine
PW_{11}	$[PW_{11}O_{39}]^{7-}$
P_2W_{18}	$K_6[P_2W_{18}O_{62}]\cdot14H_2O$
P_5W_{29}	$[NaP_5W_{29}O_{107}]^{14-}$
P_5W_{30}	$[NaP_5W_{30}O_{110}]^{14-}$
$PW_{10}Ti_2$	$[PW_{10}Ti_2O_{40}]^{7-}$
S180	Murine sarcoma cells
SARS-V	Severe acute respiratory syndrome virus
SARS-CoV	Severe acute respiratory syndrome coronavirus
SERCA	Sarco(endo)plasmatic membrane calcium ATPase
$SiW_9Nb_3O_{40}$	$Cs_2K_4Na[SiW_9Nb_3O_{40}]$
$SiVW_{11}$	$[SiVW_{11}O_{40}]^{7-}$
Sb_9W_{21}	$[KSb_9W_{21}O_{86}]^{18-}$
SR	Survival rate
TeW_6	$[TeW_6O_{24}]^{6-}$

U-87	Human brain-like glioblastoma cells
U251	Human malignant glioma cells
VRSA	Vancomycin-resistant *Staphylococcus aureus*
V_{10}	$[V_{10}O_{28}]^{6-}$
$V_{18}O_{42}$	$[V_{18}O_{42}(H_2O)]^{12-}$

References

1. Bijelic, A.; Aureliano, M.; Rompel, A. The antibacterial activity of polyoxometalates: Structures, antibiotic effects and future perspectives. *Chem. Commun.* **2018**, *54*, 1153–1169. [CrossRef] [PubMed]
2. Bijelic, A.; Aureliano, M.; Rompel, A. Polyoxometalates as Potential Next-Generation Metallodrugs in the Combat against Cancer. *Angew. Chem. Int. Ed.* **2019**, *58*, 2980–2999. [CrossRef] [PubMed]
3. Hayashi, Y. Hetero and lacunary polyoxovanadate chemistry: Synthesis, reactivity and structural aspects. *Coord. Chem. Rev.* **2011**, *255*, 2270–2280. [CrossRef]
4. Chen, X.; Yan, S.; Wang, H.; Hu, Z.; Wan, X.; Huo, M. Aerobic oxidation of starch catalyzed by isopolyoxovanadate $Na_4Co(H_2O)_6V_{10}O_{28}$. *Carbohydr. Polym.* **2015**, *117*, 673–680. [CrossRef] [PubMed]
5. Mohapatra, L.; Parida, K.M. Dramatic activities of vanadate intercalated bismuth doped LDH for solar light photocatalysis. *Phys. Chem. Chem. Phys.* **2014**, *16*, 16985–16996. [CrossRef] [PubMed]
6. Bijelic, A.; Rompel, A. The use of polyoxometalates in protein crystallography—An attempt to widen a well-known bottleneck. *Coord. Chem. Rev.* **2015**, *299*, 22–38. [CrossRef]
7. Zhao, W.; Wang, C.; Dong, S.; Li, Y.; Zhang, D.; Han, L.W. In vitro study on the antitumor activity of several new polyoxometalates. In Proceedings of the 2011 International Conference on Human Health and Biomedical Engineering, Jilin, China, 19–22 August 2011; pp. 594–597.
8. Wang, L.; Yu, K.; Zhou, B.-B.; Su, Z.-H.; Gao, S.; Chu, L.-L.; Liu, J.-R.; Wang, L. The inhibitory effects of a new cobalt-based polyoxometalate on the growth of human cancer cells. *Dalt. Trans.* **2014**, *43*, 6070. [CrossRef] [PubMed]
9. Li, Y.-T.; Zhu, C.-Y.; Wu, Z.-Y.; Jiang, M.; Yan, C.-W. Synthesis, crystal structures and anticancer activities of two decavanadate compounds. *Transit. Met. Chem.* **2010**, *35*, 597–603. [CrossRef]
10. Zhang, B.; Qiu, J.; Wu, C.; Li, Y.; Liu, Z. Anti-tumor and immunomodulatory activity of iron hepta-tungsten phosphate oxygen clusters complex. *Int. Immunopharmacol.* **2015**, *29*, 293–301. [CrossRef]
11. Dianat, S.; Bordbar, A.K.; Tangestaninejad, S.; Yadollahi, B.; Zarkesh-Esfahani, S.H.; Habibi, P. ctDNA binding affinity and in vitro antitumor activity of three Keggin type polyoxotungstates. *J. Photochem. Photobiol. B Biol.* **2013**, *124*, 27–33. [CrossRef] [PubMed]
12. Fu, L.; Gao, H.; Yan, M.; Li, S.; Li, X.; Dai, Z.; Liu, S. Polyoxometalate-Based Organic-Inorganic Hybrids as Antitumor Drugs. *Small* **2015**, *11*, 2938–2945. [CrossRef]
13. Bâlici, Ş.; Şuşman, S.; Rusu, D.; Nicula, G.Z.; Soriţău, O.; Rusu, M.; Biris, A.S.; Matei, H. Differentiation of stem cells into insulin-producing cells under the influence of nanostructural polyoxometalates. *J. Appl. Toxicol.* **2016**, *36*, 373–384. [CrossRef] [PubMed]
14. Lee, S.Y.; Fiene, A.; Li, W.; Hanck, T.; Brylev, K.A.; Fedorov, V.E.; Lecka, J.; Haider, A.; Pietzsch, H.J.; Zimmermann, H.; et al. Polyoxometalates—Potent and selective ecto-nucleotidase inhibitors. *Biochem Pharmacol.* **2015**, *93*, 171–181. [CrossRef] [PubMed]
15. Saeed, S.H.; Al-Oweini, R.; Haider, A.; Kortz, U.; Iqbal, J. Cytotoxicity and enzyme inhibition studies of polyoxometalates and their chitosan nanoassemblies. *Toxicol. Rep.* **2014**, *1*, 341–352.
16. Iqbal, J.; Barsukova-Stuckart, M.; Ibrahim, M.; Ali, S.U.; Khan, A.A.; Kortz, U. Polyoxometalates as potent inhibitors for acetyl and butyrylcholinesterases and as potential drugs for the treatment of Alzheimer's disease. *Med. Chem. Res.* **2012**, *22*, 1224–1228. [CrossRef]
17. Stephan, H.; Kubeil, M.; Emmerling, F.; Müller, C.E. Polyoxometalates as Versatile Enzyme Inhibitors. *Eur. J. Inorg. Chem.* **2012**, *2013*, 1585–1594. [CrossRef]
18. Turner, T.L.; Nguyen, V.H.; McLauchlan, C.C.; Dymon, Z.; Dorsey, B.M.; Hooker, J.D.; Jones, M. Inhibitory effects of decavanadate on several enzymes and Leishmania tarentolae In Vitro. *J. Inorg. Biochem.* **2012**, *108*, 96–104. [CrossRef] [PubMed]
19. Sanseverino, I.; Navarro-Cuenca, A.; Loos, R.; Marinov, D.; Lettieri, T. *State of the Art on the Contribution of Water to Antimicrobial Resistance*; JRC Technical Report; Publications Office of the European Union: Luxembourg, 2018. [CrossRef]
20. Rodriguez-Mozaz, S.; Vaz-Moreira, I.; Della Giustina, S.V.; Llorca, M.; Barceló, D.; Schubert, S.; Berendonk, T.U.; Michael-Kordatou, I.; Fatta-Kassinos, D.; Martinez, J.L.; et al. Antibiotic residues in final effluents of European wastewater treatment plants and their impact on the aquatic environment. *Environ. Int.* **2020**, *140*, 105733. [CrossRef] [PubMed]
21. Gomez Cortes, L.; Marinov, D.; Sanseverino, I.; Navarro-Cuenca, A.; Niegowska, M.; Porcel Rodriguez, E.; Lettieri, T. *Selection of Substances for the 3rd Watch List under the Water Framework Directive*; JRC Technical Report; Publications Office of the European Union: Luxembourg, 2020. [CrossRef]
22. Rhule, J.; Hill, C.; Judd, D.; Schinazi, R. Polyoxometalates in Medicine. *Chem. Rev.* **1998**, *98*, 327–358. [CrossRef] [PubMed]
23. Hasenknopf, B. Polyoxometalates: Introduction to a class of inorganic compounds and their biomedical applications. *Front. Biosci.* **2005**, *10*, 275. [CrossRef] [PubMed]
24. Yamase, T. Anti-tumor, -viral, and -bacterial activities of polyoxometalates for realizing an inorganic drug. *J. Mater. Chem.* **2005**, *15*, 4773. [CrossRef]

25. Sivakumar, R.; Thomas, J.; Yoon, M. Polyoxometalate-Based Molecular/Nano Composites: Advances in Environmental Remediation by Photocatalysis and Biomimetic Approaches to Solar Energy Conversion. *J. Photochem. Photobiol. C* **2012**, *13*, 277–298. [CrossRef]

26. Omwoma, S.; Gore, C.T.; Ji, Y.; Hu, C.; Song, Y.F. Environmentally Benign Polyoxometalate Materials. *Coord. Chem. Rev.* **2015**, *286*, 17–29. [CrossRef]

27. Lai, S.Y.; Ng, K.H.; Cheng, C.K.; Nur, H.; Nurhadi, M.; Arumugam, M. Photocatalytic Remediation of Organic Waste over Keggin-Based Polyoxometalate Materials: A Review. *Chemosphere* **2021**, *263*, 128244. [CrossRef] [PubMed]

28. He, R.; Xue, K.; Wang, J.; Yan, Y.; Peng, Y.; Yang, T.; Hu, Y.; Wang, W. Nitrogen-deficient g-C$_3$N$_x$/POMs porous nanosheets with P–N heterojunctions capable of the efficient photocatalytic degradation of ciprofloxacin. *Chemosphere* **2020**, *259*, 127465. [CrossRef]

29. Shi, H.; Zhao, T.; Wang, J.; Wang, Y.; Chen, Z.; Liu, B.; Ji, H.; Wang, W.; Zhang, G.; Li, Y. Fabrication of g-C3N4/PW12/TiO2 composite with significantly enhanced photocatalytic performance under visible light. *J. Alloy. Compd.* **2021**, *860*, 157924. [CrossRef]

30. Cheng, P.; Wang, Y.; Sarakha, M.; Mailhot, G. Enhancement of the photocatalytic activity of decatungstate, W10O324−, for the oxidation of sulfasalazine/sulfapyridine in the presence of hydrogen peroxide. *J. Photochem. Photobiol. A Chem.* **2021**, *404*, 112890. [CrossRef]

31. Silva, E.S.D.; Sarakha, M.; Burrows, H.D.; Wong-Wah-Chung, P. Decatungstate anion as an efficient photocatalytic species for the transformation of the pesticide 2-(1-naphthyl)acetamide in aqueous solution. *J. Photochem. Photobiol. A* **2017**, *334*, 61–73. [CrossRef]

32. Li, G.; Zhang, K.; Li, C.; Gao, R.; Cheng, Y.; Hou, L.; Wang, Y. Solvent-free method to encapsulate polyoxometalate into metal-organic frameworks as efficient and recyclable photocatalyst for harmful sulfamethazine degrading in water. *Appl. Catal. B Environ.* **2019**, *245*, 753–759. [CrossRef]

33. López, Y.C.; Viltres, H.; Gupta, N.K.; Acevedo-Peña, P.; Leyva, C.; Ghaffari, Y.; Gupta, A.; Kim, S.; Bae, J.; Kim, K.S. Transition metal-based metal–organic frameworks for environmental applications: A review. *Env. Chem. Lett.* **2021**, *19*, 1295–1334. [CrossRef]

34. Majdafshar, M.; Piryaei, M.; Abolghasemi, M.M.; Rafiee, E. Polyoxometalate-based ionic liquid coating for solid phase microextraction of triazole pesticides in water samples. *Sep. Sci. Technol.* **2019**, *54*, 1553–1559. [CrossRef]

35. Martinetto, Y.; Pegot, B.; Roch-Marchal, C.; Cottyn-Boitte, B.; Floquet, S. Designing functional polyoxometalate-based ionic liquid crystals and ionic liquids. *Eur. J. Inorg. Chem.* **2020**, *2020*, 228–247. [CrossRef]

36. Misra, A.; Zambrzycki, C.; Kloker, G.; Kotyrba, A.; Anjass, M.H.; Castillo, I.F.; Mitchell, S.G.; Güttel, R.; Streb, C. Water Purification and Microplastics Removal Using Magnetic Polyoxometalate-Supported Ionic Liquid Phases (magPOM-SILPs). *Angew. Chem. Int. Ed.* **2020**, *59*, 1601–1605. [CrossRef]

37. Worthington, R.J.; Melander, C. Combination approaches to combat multidrug-resistant bacteria. *Trends Biotechnol.* **2013**, *31*, 177–184. [CrossRef]

38. Tajima, Y.; Nagasawa, Z.; Tadano, J. A factor found in aged tungstate solution enhanced the antibacterial effect of beta-lactams on methicillin-resistant Staphylococcus aureus. *Microbiol. Immunol.* **1993**, *37*, 695–703. [CrossRef] [PubMed]

39. Raynaud, M.; Chermann, J.C.; Plata, F.; Mathé, G. Inhibitors of viruses of the leukemia and murine sarcoma group. Biological inhibitor CJMR. *Comptes Rendus Hebd. Seances L'academie des Sci. Ser. D* **1971**, *272*, 2038–2040.

40. Mukherjee, H.N. Treatment of Cancer of the Intestinal Tract with a Complex Compound of Phosphotungstic Phosphomolybdic Acids and Caffeine. *J. Indian Med. Assoc.* **1965**, *44*, 477–479. [PubMed]

41. Ling, L.L.; Schneider, T.; Peoples, A.J.; Spoering, A.L.; Engels, I.; Conlon, B.P.; Mueller, A.; Schäberle, T.F.; Hughes, D.E.; Epstein, S.; et al. A new antibiotic kills pathogens without detectable resistance. *Nature* **2015**, *517*, 455–459. [CrossRef] [PubMed]

42. Fiers, W.D.; Craighead, M.; Singh, I. Teixobactin and Its Analogues: A New Hope in Antibiotic Discovery. *ACS Infect. Dis.* **2017**, *3*, 688–690. [CrossRef]

43. Dizaj, S.M.; Lotfipour, F.; Barzegar-Jalali, M.; Zarrintan, M.H.; Adibkia, K. Antimicrobial activity of the metals and metal oxide nanoparticles. *Mater. Sci. Eng. C* **2014**, *44*, 278–284. [CrossRef]

44. Farzana, R.; Iqra, P.; Hunaiza, T. Antioxidant and antimicrobial effects of polyoxometalates. *Microbiol. Curr. Res.* **2018**, *2*, 7–11.

45. Misra, A.; Franco Castillo, I.; Müller, D.P.; González, C.; Eyssautier-Chuine, S.; Ziegler, A.; de laFuente, J.M.; Mitchell, S.G.; Streb, C. Polyoxometalate-Ionic Liquids (POM-ILs) as Anticorrosion and Antibacterial Coatings for Natural Stones. *Angew. Chem. Int. Ed.* **2018**, *57*, 14926–14931. [CrossRef] [PubMed]

46. Olsen, M.R.; Colliard, I.; Rahman, T.; Miyaishi, T.C.; Harper, B.; Harper, S.; Nyman, M. Hybrid Polyoxometalate Salt Adhesion by Butyltin Functionalization. *ACS Appl Mater Interfaces* **2021**, *13*, 19497–19506. [CrossRef] [PubMed]

47. Rajkowska, K.; Koziróg, A.; Otlewska, A.; Piotrowska, M.; Atrián-Blasco, E.; Franco-Castillo, I.; Mitchell, S.G. Antifungal Activity of Polyoxometalate-Ionic Liquids on Historical Brick. *Molecules* **2020**, *25*, 5663. [CrossRef] [PubMed]

48. Chen, K.; Liu, Y.; Li, M.; Liu, L.; Yu, Q.; Wu, L. Amelioration of enteric dysbiosis by polyoxotungstates in mice gut. *J. Inorg. Biochem.* **2022**, *226*, 111654. [CrossRef] [PubMed]

49. Chen, K.; Yu, Q.; Liu, Y.; Yin, P. Bacterial hyperpolarization modulated by polyoxometalates for solutions of antibiotic resistance. *J. Inorg. Biochem.* **2021**, *220*, 111463. [CrossRef] [PubMed]

50. Xu, Z.; Chen, K.; Li, M.; Hu, C.; Yin, P. Sustained release of Ag$^+$ confined inside polyoxometalates for long-lasting bacterial resistance. *Chem. Commun.* **2020**, *56*, 5287–5290. [CrossRef]

51. Gonzalez, A.; Galvez, N.; Clemente-Leon, M.; Dominguez-Vera, J.M. Electrochromic polyoxometalate material as a sensor of bacterial activity. *Chem. Commun.* **2015**, *51*, 10119–10122. [CrossRef]

52. Chen, S.; Wu, G.; Long, D.; Liu, Y. Preparation, characterization and antibacterial activity of chitosan-$Ca_3V_{10}O_{28}$ complex membrane. *Carbohydr. Polym.* **2006**, *64*, 92–97. [CrossRef]

53. Yamase, T.; Fukuda, N.; Tajima, Y. Synergistic effect of polyoxotungstates in combination with beta-lactam antibiotics on antibacterial activity against methicillin-resistant Staphylococcus aureus. *Biol. Pharm. Bull.* **1996**, *19*, 459–465. [CrossRef]

54. Balici, S.; Niculae, M.; Pall, E.; Rusu, M.; Rusu, D.; Matei, H. Antibiotic-Like Behaviour of Polyoxometalates. In vitro comparative study: Seven polyoxotungstates-nine antibiotics against gram-positive and gram-negative bacteria. *Rev. Chim.* **2016**, *67*, 485–490.

55. Fang, Y.; Xing, C.; Liu, J.; Zhang, Y.; Li, M.; Han, Q. Supermolecular film crosslinked by polyoxometalate and chitosan with superior antimicrobial effect. *Int. J. Biol. Macromol.* **2020**, *154*, 732–738. [CrossRef]

56. Fukuda, N.; Yamase, T.; Tajima, Y. Inhibitory effect of polyoxotungstates on the production of penicillin- binding proteins and β-lactamase against methicillin-resistant Staphylococcus aureus. *Biol. Pharm. Bull.* **1999**, *22*, 463–470. [CrossRef] [PubMed]

57. Gumerova, N.; Al-Sayed, E.; Krivosudský, L.; Čipčić-Paljetak, H.; Verbanac, D.; Rompel, A. Antibacterial Activity of Polyoxometalates Against Moraxella catarrhalis. *Front. Chem.* **2018**, *6*, 336–345. [CrossRef]

58. Inoue, M.; Segawa, K.; Matsunaga, S.; Matsumoto, N.; Oda, M.; Yamase, T. Antibacterial activity of highly negative charged polyoxotungstates, $K_{27}[KAs_4W_{40}O_{140}]$ and $K_{18}[KSb_9W_{21}O_{86}]$, and Keggin-structural polyoxotungstates against Helicobacter pylori. *J. Inorg. Biochem.* **2005**, *99*, 1023–1031. [CrossRef] [PubMed]

59. Inoue, M.; Suzuki, T.; Fujita, Y.; Oda, M.; Matsumoto, N.; Yamase, T. Enhancement of antibacterial activity of beta-lactam antibiotics by $P_2W_{18}O_{62}]^{6-}$, $[SiMo_{12}O_{40}]^{4-}$, and $[PTi_2W_{10}O_{40}]^{7-}$ against methicillin-resistant and vancomycin-resistant Staphylococcus aureus. *J. Inorg. Biochem.* **2006**, *100*, 1225–1233. [CrossRef]

60. Samart, N.; Arhouma, Z.; Kumar, S.; Murakami, H.A.; Crick, D.C.; Crans, D.C. Decavanadate inhibits microbacterial growth more potently than other oxovanadates. *Front. Chem.* **2018**, *6*, 519. [CrossRef] [PubMed]

61. Missina, J.M.; Gavinho, B.; Postal, K.; Santana, F.S.; Valdameri, G.; De Souza, E.M.; Hughes, D.L.; Ramirez, M.I.; Soares, J.F.; Nunes, G.G. Effects of Decavanadate Salts with Organic and Inorganic Cations on Escherichia coli, Giardia intestinalis, and Vero Cells. *Inorg. Chem.* **2018**, *57*, 11930–11941. [CrossRef]

62. Marques-Da-Silva, D.; Fraqueza, G.; Lagoa, R.; Vannathan, A.A.; Mal, S.S.; Aureliano, M. Polyoxovanadate inhibition of: Escherichia coli growth shows a reverse correlation with Ca^{2+}-ATPase inhibition. *New J. Chem.* **2019**, *43*, 17577–17587. [CrossRef]

63. Fukuda, N.; Yamase, T. In Vitro Antibacterial Activity of Vanadate and Vanadyl Compounds against Streptococcus pneumoniae. *Biol. Pharm. Bull.* **1997**, *20*, 927–930. [CrossRef]

64. Yang, P.; Bassil, B.S.; Lin, Z.; Haider, A.; Alfaro-Espinoza, G.; Ullrich, M.S.; Silvestru, C.; Kortz, U. Organoantimony(III)-Containing Tungstoarsenates(III): From Controlled Assembly to Biological Activity. *Chem. Eur. J.* **2015**, *21*, 15600–15606. [CrossRef] [PubMed]

65. Yang, P.; Lin, Z.; Bassil, B.S.; Alfaro-Espinoza, G.; Ullrich, M.S.; Li, M.X.; Silvestru, C.; Kortz, U. Tetra-Antimony(III)-Bridged 18-Tungsto-2-Arsenates(V), [(LSbIII)4(A-α-AsVW9O34)2]10– (L = Ph, OH): Turning Bioactivity On and Off by Ligand Substitution. *Inorg. Chem.* **2016**, *55*, 3718–3720. [CrossRef] [PubMed]

66. Liu, H.; Zou, Y.L.; Zhang, L.; Liu, J.X.; Song, C.Y.; Chai, D.F.; Gao, G.G.; Qiu, Y.F. Polyoxometalate cobalt–gatifloxacin complex with DNA binding and antibacterial activity. *J. Coord. Chem.* **2014**, *67*, 2257–2270. [CrossRef]

67. Yang, F.-C.; Wu, K.-H.; Lin, W.-P.; Hu, M.-K. Preparation and antibacterial efficacy of bamboo charcoal/polyoxometalate biological protective material. *Microporous Mesoporous Mater.* **2009**, *118*, 467–472. [CrossRef]

68. Wu, K.H.; Yu, P.Y.; Yang, C.C.; Wang, G.P.; Chao, C.M. Preparation and characterization of polyoxometalate-modified poly(vinyl alcohol)/polyethyleneimine hybrids as a chemical and biological self-detoxifying material. *Polym. Degrad. Stab.* **2009**, *94*, 1411–1418. [CrossRef] [PubMed]

69. Kubo, A.-L.; Kremer, L.; Herrmann, S.; Mitchell, S.G.; Bondarenko, O.M.; Kahru, A.; Streb, C. Antimicrobial Activity of Polyoxometalate Ionic Liquids against Clinically Relevant Pathogens. *ChemPlusChem* **2017**, *82*, 867–871. [CrossRef]

70. Fraqueza, G.; Ohlin, C.A.; Casey, W.H.; Aureliano, M. Sarcoplasmic reticulum calcium ATPase interactions with decaniobate, decavanadate, vanadate, tungstate and molybdate. *J. Inorg. Biochem.* **2012**, *107*, 82–89. [CrossRef]

71. Fraqueza, G.; Carvalho, L.A.; de Marques, E.B.; Marques, M.P.M.; Maia, L.; Ohlin, C.A.; Casey, W.H.; Aureliano, M. Decavanadate, decaniobate, tungstate and molybdate interactions with sarcoplasmic reticulum Ca^{2+}-ATPase: Quercetin prevents cysteine oxidation by vanadate but does not reverse ATPase inhibition. *Dalt. Trans.* **2012**, *41*, 12749. [CrossRef]

72. Gumerova, N.; Krivosudský, L.; Fraqueza, G.; Breibeck, J.; Al-Sayed, E.; Tanuhadi, E.; Bijelic, A.; Fuentes, J.; Aureliano, M.; Rompel, A. The P-type ATPase inhibiting potential of polyoxotungstates. *Metallomics* **2018**, *10*, 287–295. [CrossRef]

73. Fraqueza, G.; Fuentes, J.; Krivosudský, L.; Dutta, S.; Mal, S.S.; Roller, A.; Giester, G.; Rompel, A.; Aureliano, M. Inhibition of Na+/K+- and Ca^{2+}-ATPase activities by phosphotetradecavanadate. *J. Inorg. Biochem.* **2019**, *197*, 110700. [CrossRef]

74. Fang, Y.; Xing, C.; Zhan, S.; Zhao, M.; Li, M.; Liu, H. A polyoxometalate-modified magnetic nanocomposite: A promising antibacterial material for water treatment. *J. Mater. Chem. B* **2019**, *7*, 1933–1944. [CrossRef] [PubMed]

75. Ma, X.; Zhou, F.; Yue, H.; Hua, J.; Ma, P. A nano-linear zinc-substituted phosphomolybdate with reactive oxygen species catalytic ability and antibacterial activity. *J. Mol. Struct.* **2019**, *1198*, 126865. [CrossRef]

76. Missina, J.M.; Leme, L.B.P.; Postal, K.; Santana, F.S.; Hughes, D.L.; de Sá, E.L.; Ribeiro, R.R.; Nunes, G.G. Accessing decavanadate chemistry with tris(hydroxymethyl)aminomethane, and evaluation of methylene blue bleaching. *Polyhedron* **2020**, *180*, 114414. [CrossRef]

77. Harduin-Lepers, A.; Mollicone, R.; Delannoy, P.; Oriol, R. The animal sialyltransferases and sialyltransferase-related genes: A phylogenetic approach. *Glycobiology* **2005**, *15*, 805–817. [CrossRef]

78. Negishi, M.; Pedersen, L.G.; Petrotchenko, E.; Shevtsov, S.; Gorokhov, A.; Kakuta, Y.; Pedersen, L.C. Structure and function of sulfotransferases. *Arch. Biochem. Biophys.* **2001**, *390*, 149–157. [CrossRef] [PubMed]

79. Seko, A.; Yamase, T.; Yamashita, K. Polyoxometalates as effective inhibitors for sialyl- and sulfotransferases. *J. Inorg. Biochem.* **2009**, *103*, 1061–1066. [CrossRef] [PubMed]

80. Dong, Z.; Tan, R.; Cao, J.; Yang, Y.; Kong, C.; Du, J.; Zhu, S.; Zhang, Y.; Lu, J.; Huang, B.; et al. Discovery of polyoxometalate-based HDAC inhibitors with profound anticancer activity in vitro and in vivo. *Eur. J. Med. Chem.* **2011**, *46*, 2477–2484. [CrossRef] [PubMed]

81. Raza, R.; Matin, A.; Sarwar, S.; Barsukova-Stuckart, M.; Ibrahim, M.; Kortz, U.; Iqbal, J. Polyoxometalates as potent and selective inhibitors of alkaline phosphatases with profound anticancer and amoebicidal activities. *Dalt. Trans.* **2012**, *41*, 14329. [CrossRef] [PubMed]

82. León, I.E.; Porro, V.; Astrada, S.; Egusquiza, M.G.; Cabello, C.I.; Bollati-Fogolin, M.; Etchevery, S.B. Polyoxometalates as antitumor agents: Bioactivity of a new polyoxometalate with copper on a human osteosarcoma model. *Chem. Biol. Interact.* **2014**, *222*, 87–96. [CrossRef]

83. She, S.; Bian, S.; Huo, R.; Chen, K.; Huang, Z.; Zhang, J.; Hao, J.; Wei, Y. Degradable Organically-Derivatized Polyoxometalate with Enhanced Activity against Glioblastoma Cell Line. *Sci. Rep.* **2016**, *6*, 33529. [CrossRef]

84. Cheng, M.; Li, N.; Wang, N.; Hu, K.; Xiao, Z.; Wu, P.; Wei, Y. Synthesis, structure and antitumor studies of a novel decavanadate complex with a wavelike two-dimensional network. *Polyhedron* **2018**, *155*, 313–319. [CrossRef]

85. Qi, W.; Zhang, B.; Qi, Y.; Guo, S.; Tian, R.; Sun, J.; Zhao, M. The Anti-Proliferation Activity and Mechanism of Action of $K_{12}[V_{18}O_{42}(H_2O)].6H_2O$ on Breast Cancer Cell Lines. *Molecules* **2018**, *22*, 1535. [CrossRef]

86. Zheng, Y.; Gan, H.; Zhao, Y.; Li, W.; Wu, Y.; Yan, X.; Wang, Y.; Li, J.; Li, J.; Wang, X. Self-Assembly and Antitumor Activity of a Polyoxovanadate-Based Coordination Nanocage. *Chem. A Eur. J.* **2019**, *25*, 15326–15332. [CrossRef]

87. Sun, T.; Cui, W.; Yan, M.; Qin, G.; Guo, W.; Gu, H.; Liu, S.; Wu, Q. Target Delivery of a Novel Antitumor Organoplatinum(IV)-Substituted Polyoxometalate Complex for Safer and More Effective Colorectal Cancer Therapy in vivo. *Adv. Mater.* **2016**, *28*, 7397–7404. [CrossRef] [PubMed]

88. Dong, Z.; Yang, Y.; Liu, S.; Lu, J.; Huang, B.; Zhang, Y. HDAC inhibitor PAC-320 induces G2/M cell cycle arrest and apoptosis in human prostate cancer. *Oncotarget* **2018**, *9*, 512–523. [CrossRef] [PubMed]

89. Yong, Y.; Zhang, C.; Gu, Z.; Du, J.; Guo, Z.; Dong, X.; Xie, J.; Zhang, G.; Liu, X.; Zhao, Y. Polyoxometalate-Based Radiosensitization Platform for Treating Hypoxic Tumors by Attenuating Radioresistance and Enhancing Radiation Response. *ACS Nano* **2017**, *11*, 7164–7176. [CrossRef] [PubMed]

90. Boulmier, A.; Feng, X.; Oms, O.; Mialane, P.; Rivière, E.; Shin, C.J.; Yao, J.; Kubo, T.; Furuta, T.; Oldfield, E.; et al. Anticancer Activity of Polyoxometalate-Bisphosphonate Complexes: Synthesis, Characterization, in vitro and in vivo Results. *Inorg. Chem.* **2017**, *56*, 7558–7565. [CrossRef]

91. Aureliano, M.; Gumerova, N.I.; Sciortino, G.; Garribba, E.; Rompel, A.; Crans, D.C. Polyoxovanadates with emerging biomedical activities. *Coord. Chem. Rev.* **2021**, *447*, 214143. [CrossRef]

92. Guedes, G.; Wang, S.; Santos, H.A.; Sousa, F.L. Polyoxometalate Composites in Cancer Therapy and Diagnostics. *Eur. J. Inorg. Chem.* **2020**, *2020*, 2121–2132. [CrossRef]

93. Guedes, G.; Wang, S.; Fontana, F.; Figueiredo, P.; Linden, J.; Correia, A.; Pinto, R.J.B.; Hietala, S.; Sousa, F.L.; Santos, H.A. Dual-Crosslinked Dynamic Hydrogel Incorporating {Mo-154} with pH and NIR Responsiveness for Chemo-Photothermal Therapy. *Adv. Mater.* **2021**, *33*, 2007761. [CrossRef]

94. Karimian, D.; Yadollahi, B.; Mirkhani, V. Dual functional hybrid-polyoxometalate as a new approach for multidrug delivery. *Microporous Mesoporous Mater.* **2017**, *247*, 23–30. [CrossRef]

95. Nienberg, C.; Garmann, C.; Gratz, A.; Bollacke, A.; Götz, C.; Jose, J. Identification of a Potent Allosteric Inhibitor of Human Protein Kinase CK2 by Bacterial Surface Display Library Screening. *Pharmaceuticals* **2017**, *10*, 6. [CrossRef]

96. Prudent, R.; Moucadel, V.; Laudet, B.; Barette, C.; Lafanechère, L.; Hasenknopf, B.; Li, J.; Bareyt, S.; Lacote, E.; Thorimbert, S.; et al. Identification of Polyoxometalates as Nanomolar Noncompetitive Inhibitors of Protein Kinase CK2. *Chem. Biol.* **2008**, *15*, 683–692. [CrossRef] [PubMed]

97. Pimpão, C.; da Silva, I.V.; Mósca, A.F.; Pinho, J.O.; Gaspar, M.M.; Gumerova, N.I.; Rompel, A.; Aureliano, M.; Soveral, G. The aquaporin-3-inhibiting potential of polyoxotungstates. *Int. J. Mol. Sci.* **2020**, *21*, 2467. [CrossRef] [PubMed]

98. Soveral, G.; Casini, A. Aquaporin modulators: A patent review (2010–2015). *Expert Opin. Ther. Pat.* **2016**, *27*, 49–62. [CrossRef]

99. Fonseca, C.; Fraqueza, G.; Carabineiro, S.A.C.; Aureliano, M. The Ca^{2+}-ATPase inhibition potential of gold (I,III) compounds. *Inorganics* **2020**, *8*, 49. [CrossRef]

100. Berrocal, M.; Cordoba-Granados, J.J.; Carabineiro, S.A.C.; Gutierrez-Merino, C.; Aureliano, M.; Mata, A.M. Gold Compounds Inhibit the Ca2+-ATPase Activity of Brain PMCA and Human Neuroblastoma SH-SY5Y Cells and Decrease Cell Viability. *Metals* **2021**, *11*, 1934. [CrossRef]

101. Zhu, N.; Zhang, D.; Wang, W.; Li, X.; Yang, B.; Song, J.; Zhao, X.; Huang, B.; Shi, W.; Lu, R.; et al. A Novel Coronavirus from Patients with Pneumonia in China, 2019. *N. Engl. J. Med.* **2020**, *382*, 727–733. [CrossRef] [PubMed]

102. Flütsch, A.; Schroeder, T.; Grütter, M.G.; Patzke, G.R. HIV-1 protease inhibition potential of functionalized polyoxometalates. *Bioorg. Med. Chem. Lett.* **2011**, *21*, 1162–1166. [CrossRef]

103. Qi, Y.; Xiang, Y.; Wang, J.; Qi, Y.; Li, J.; Niu, J.; Zhong, J. Inhibition of hepatitis C virus infection by polyoxometalates. *Antivir. Res.* **2013**, *100*, 392–398. [CrossRef]

104. Witvrouw, M.; Weigold, H.; Pannecouque, C.; Schols, D.; De Clercq, E.; Holan, G. Potent anti-HIV (type 1 and type 2) activity of polyoxometalates: Structure-activity relationship and mechanism of action. *J. Med. Chem.* **2000**, *43*, 778–783. [CrossRef] [PubMed]

105. Francese, R.; Civra, A.; Rittà, M.; Donalisio, M.; Argenziano, M.; Cavalli, R.; Mougharbel, A.; Kortz, U.; Lembo, D. Anti-zika virus activity of polyoxometalates. *Antivir. Res.* **2019**, *163*, 29–33. [CrossRef] [PubMed]

106. Wang, J.; Liu, Y.; Xu, K.; Qi, Y.; Zhong, J.; Zhang, K.; Li, J.; Wang, E.; Wu, Z.; Kan, G.Z. Broad-spectrum antiviral property of polyoxometalate localized on a cell surface. *ACS Appl. Mater. Interfaces* **2014**, *6*, 9785–9789. [CrossRef]

107. Wang, X.; Wang, J.; Zhang, W.; Li, B.; Zhu, Y.; Hu, Q.; Yang, Y.; Zhang, X.; Yan, H.; Zeng, Y. Inhibition of human immunodeficiency virus type 1 entry by a keggin polyoxometalate. *Viruses* **2018**, *10*, 265. [CrossRef] [PubMed]

108. Li, Q.; Zhang, H.; Qi, Y.; Wang, J.; Li, J.; Niu, J. Antiviral effects of a niobium-substituted heteropolytungstate on hepatitis B virus-transgenic mice. *Drug Dev. Res.* **2019**, *80*, 1062–1070. [CrossRef] [PubMed]

109. Zhang, H.; Qi, Y.; Ding, Y.; Wang, J.; Li, Q.; Zhang, J.; Jiang, Y.; Chi, X.; Li, J.; Niu, J. Synthesis, characterization and biological activity of a niobium-substituted-heteropolytungstate on hepatitis B virus. *Bioorg. Med. Chem. Lett.* **2012**, *22*, 1664–1669. [CrossRef] [PubMed]

110. Hosseini, S.M.; Amini, E.; Kheiri, M.T.; Mehrbod, P.; Shahidi, M.; Zabihi, E. Anti-influenza Activity of a Novel Polyoxometalate Derivative (POM-4960). *Int. J. Mol. Cell. Med.* **2012**, *1*, 21–29.

111. Shigeta, S.; Mori, S.; Yamase, T.; Yamamoto, N.; Yamamoto, N. Anti-RNA virus activity of polyoxometalates. *Biomed. Pharmacother.* **2006**, *60*, 211–219. [CrossRef]

112. Chanh, C.T.; Dreesman, G.R.; Kennedy, R.C. Monoclonal anti-idiotypic antibody mimics the CD4 receptor and binds human immunodeficiency virus. *Proc. Natl. Acad. Sci. USA* **1987**, *84*, 3891–3895. [CrossRef]

113. Douglas, T.; Young, M. Host-guest encapsulation of materials by assembled virus protein cages. *Nature* **1998**, *393*, 152–155. [CrossRef]

114. Yamase, T. Polyoxometalates Active against Tumors, Viruses, and Bacteria. *Prog. Mol. Subcell. Biol.* **2013**, *54*, 65–116. [CrossRef]

115. Liu, J.; Mei, W.; Li, Y.; Wang, E.; Ji, L.; Tao, P. Antiviral Activity of Mixed-Valence Rare Earth Borotungstate Heteropoly Blues against Influenza Virus in Mice. *Antivir. Chem. Chemother.* **2000**, *11*, 367–372. [CrossRef]

116. Shao, C.; Wang, J.-P.; Yang, G.-C.; Su, Z.-M.; Hu, D.-H.; Sun, C.-C. Interactions of $[Mo_6O_{19}]^{2-}$ and its derivatives substituted with organic groups inhibitor with SARS-CoV 3CLpro by molecular modeling. *Gaodeng Xuexiao Huaxue Xuebao/Chem. J. Chin. Univ.* **2008**, *29*, 165–169.

117. Hu, D.; Shao, C.; Guan, W.; Su, Z.; Sun, J. Studies on the interactions of Ti-containing polyoxometalates (POMs) with SARS-CoV 3CLpro by molecular modeling. *J. Inorg. Biochem.* **2007**, *101*, 89–94. [CrossRef]

118. Aureliano, M.; Ohlin, C.A.; Vieira, M.O.; Marques, M.P.M.; Casey, W.H.; Batista de Carvalho, L.A.E. Characterization of decavanadate and decaniobate solutions by Raman spectroscopy. *Dalton Trans.* **2016**, *45*, 7391–7399. [CrossRef]

119. Gumerova, N.I.; Rompel, A. Polyoxometalates in solution: Speciation under spotlight. *Chem. Soc. Rev.* **2020**, *49*, 7568–7601. [CrossRef]

120. Soares, S.S.; Henao, F.; Aureliano, M.; Gutiérrez-Merino, C. Vanadate-induced necrotic death in neonatal rat cardiomyocytes through mitochondrial membrane depolarization. *Chem. Res. Toxicol.* **2008**, *21*, 607–618. [CrossRef] [PubMed]

121. Gândara, R.M.C.; Soares, S.S.; Martins, H.; Gutiérrez-Merino, C.; Aureliano, M. Vanadate oligomers: In vivo effects in hepatic vanadium accumulation and stress markers. *J. Inorg. Biochem.* **2005**, *99*, 1238–1244. [CrossRef] [PubMed]

122. Aureliano, M.; Gumerova, N.I.; Sciortino, G.; Garribba, E.; McLauchlan, C.C.; Rompel, A.; Crans, D.C. Polyoxovanadates interactions with proteins: An overview. *Coord. Chem. Rev.* **2022**, *454*, 214344. [CrossRef]

123. Soares, S.S.; Gutiérrez-Merino, C.; Aureliano, M. Decavanadate induces mitochondrial membrane depolarization and inhibits oxygen consumption. *J. Inorg. Biochem.* **2007**, *101*, 789–796. [CrossRef] [PubMed]

124. Ramos, S.; Manuel, M.; Tiago, T.; Duarte, R.O.; Martins, J.; Gutiérrez-Merino, C.; Moura, J.J.G.; Aureliano, M. Decavanadate interactions with actin inhibit G-actin polymerization and stabilize decameric vanadate species. *J. Inorg. Biochem.* **2006**, *100*, 1734–1743. [CrossRef]

125. Aureliano, M. Decavanadate: A journey in a search of a role. *Dalton Trans.* **2009**, *42*, 9093–9100. [CrossRef] [PubMed]

126. Aureliano, M.; Fraqueza, G.; Ohlin, C.A. Ion pumps as biological targets for decavanadate. *Dalton Trans.* **2013**, *42*, 11770–11777. [CrossRef] [PubMed]

127. Crans, D.C.; Smee, J.J.; Gaidamauskas, E.; Yang, L. The chemistry and biochemistry of vanadium and the biological activities exerted by vanadium compounds. *Chem. Rev.* **2004**, *104*, 849–902. [CrossRef]

128. Al-Qatati, A.; Fontes, F.L.; Barisas, B.G.; Zhang, D.; Roess, D.A.; Crans, D.C. Raft localization of Type I Fcε receptor and degranulation of RBL-2H3 cells exposed to decavanadate, a structural model for V_2O_5. *Dalton Trans.* **2013**, *42*, 11912–11920. [CrossRef] [PubMed]

129. Aureliano, M.; Crans, D.C. Decavanadate ($V_{10}O_{28}{}^{6-}$) and oxovanadates: Oxometalates with many biological activities. *J. Inorg. Biochem.* **2009**, *103*, 536–546. [CrossRef]

130. Sciortino, G.; Aureliano, M.; Garribba, E. Rationalizing the decavanadate(V) and oxidovanadium(IV) binding to G-actin and the competition with decaniobate(V) and ATP. *Inorg. Chem.* **2021**, *60*, 334–344. [CrossRef]

19

 **BioChem**

Review

Heme-Based Gas Sensors in Nature and Their Chemical and Biotechnological Applications

Ana Claudia Silva Gondim, Wellinson Gadelha Guimarães and Eduardo Henrique Silva Sousa *

Bioinorganic Group, Department of Organic and Inorganic Chemistry, Federal University of Ceara, Fortaleza 60440-900, Brazil; acgondim@dqoi.ufc.br (A.C.S.G.); wellinson000@yahoo.com.br (W.G.G.)
* Correspondence: eduardohss@dqoi.ufc.br

Abstract: Sensing is an essential feature of life, where many systems have been developed. Diatomic molecules such as O_2, NO and CO exhibit an important role in life, which requires specialized sensors. Among the sensors discovered, heme-based gas sensors compose the largest group with at least eight different families. This large variety of proteins also exhibits many distinct ways of sensing diatomic molecules and promote a response for biological adaptation. Here, we briefly describe a story of two impressive systems of heme-based oxygen sensors, FixL from *Rhizobium* and DevS(DosS)/DosT from *Mycobacterium tuberculosis*. Beyond this, we also examined many applications that have emerged. These heme-based gas sensors have been manipulated to function as chemical and biochemical analytical systems to detect small molecules (O_2, CO, NO, CN^-), fluorophores for imaging and bioanalysis, regulation of processes in synthetic biology and preparation of biocatalysts among others. These exciting features show the robustness of this field and multiple opportunities ahead besides the advances in the fundamental understanding of their molecular functioning.

Keywords: heme-based gas sensors; DevS/DosT/DevR; FixL; bioanalytical tool; biotechnological applications

Citation: Gondim, A.C.S.; Guimarães, W.G.; Sousa, E.H.S. Heme-Based Gas Sensors in Nature and Their Chemical and Biotechnological Applications. *BioChem* **2022**, *2*, 43–63. https://doi.org/10.3390/biochem2010004

Academic Editors: Manuel Aureliano, M. Leonor Cancela, Célia M. Antunes and Ana Rodrigues Costa

Received: 20 December 2021
Accepted: 29 January 2022
Published: 8 February 2022

Publisher's Note: MDPI stays neutral with regard to jurisdictional claims in published maps and institutional affiliations.

1. Introduction

Life is a complex web of connections, where no single organism lives in complete isolation. Chemical ecology is an obvious and essential feature that emerges from this, where a multitude of molecules are produced, released, absorbed and transformed all the time. Some of them can be very important while others are extremely harmful, which sometimes requires very sophisticated alert systems. The ability to sense and respond to some molecules is a way to adapt to environmental changes, a key feature of life. Among the various molecules required to be sensed, there are very simple and small ones such as diatomic molecules that can play critical roles in life. Molecules such as H_2, N_2, O_2, NO and CO have many biological uses, and sensing systems have been found for most of them, regulating their processes [1–4].

Nature has developed exciting proteins capable of selectively binding and recognizing diatomic gases, distinguishing one single atom of oxygen from nitrogen or carbon in the triad of gases: O_2, NO and CO [2,5]. Their chemical properties as tiny gaseous molecules make it harder for proteins to bind them [1]. Indeed, most of the protein sensors for these gaseous molecules employ metal centers that are capable of making a direct bond with them. Actually, there are many highly stable metal-based compounds containing one or more CO or NO as ligands, including in metalloprotein sites (see hydrogenases, nitrogenases [6]). The use of metal ions, particularly of iron in proteins, allows sensors to interact more efficiently with O_2, NO or CO. Indeed, their binding strength can vary widely (from picomolar to micromolar), opening opportunities to probe a large range of concentrations. The most common nature of these iron-based sensing units is mononuclear iron, binuclear iron, iron-sulfur cluster or heme sites [1]. These four distinct types of iron-based sites have been found in many systems such as NO sensor NorR from *Escherichia coli* (mononuclear iron site) [7], O_2 sensor *VcBhr-DGC* from *Vibrio cholerae* (binuclear iron site, hemerythrin-like) [8], O_2 sensor FNR from *Escherichia coli* (iron–sulfur cluster site [Fe$_4$S$_4$]) [9] and O_2 sensor FixL from *Rhizobium* (heme site) [10] (Figure 1). Regarding their structural organization, these metal-based sensing units have been found within a certain region of the protein, in a different domain or as a completely independent protein (Figure 1). This organizations can be exemplified by three metalloprotein gas sensors:

1. Single domain: Whib3, an NO/O_2 sensor from *Mycobacterium tuberculosis*, exhibits an iron–sulfur cluster site (sensing unit) and a DNA-binding region (responsive unit) within the same protein region (Figure 1) [9].
2. Different domains: FixL, an O_2 sensor from *Rhizobia* bacteria, contains a heme site (sensing unit) and a histidine kinase region (responsive unit) in different domains (Figure 1) [3].
3. Independent proteins: *PaNosP*, an NO sensor from *Pseudomonas aeruginosa*, presents a heme-containing protein (sensing unit) that functions by regulating another independent histidine kinase protein (responsive unit) (Figure 1) [11].

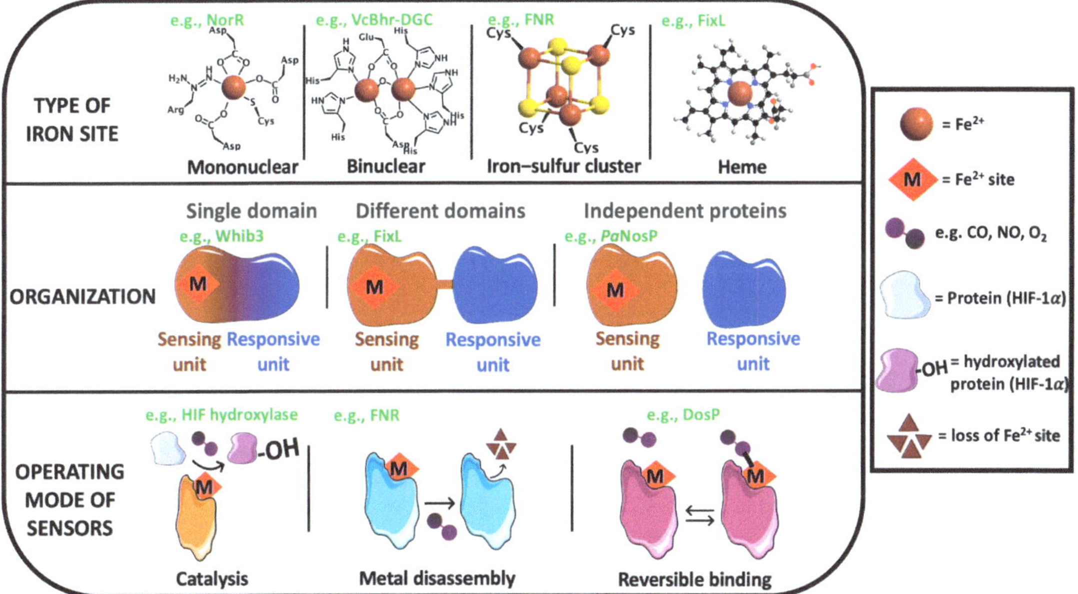

Figure 1. Basic building blocks of metal-based sensor proteins for O_2, NO and CO and molecular functioning. Firstly, four types of iron sites are briefly illustrated with examples of gas sensors: mononuclear iron site of the NO sensor NorR from *E. coli*, binuclear iron site of the O_2 sensor Bhr-DGC from *Vibrio cholerae*, iron–sulfur cluster of the O_2 sensor FNR from *E. coli*, and heme iron site of the O_2 sensor FixL from *Rhizobium*. Then, it shows the overall organization of metal-based gas sensors with three distinct structural features along with examples such as Whib3 from *Mycobacterium tuberculosis* as a single protein sensor of NO/O_2, FixL as a multi-domain O_2 sensor and *PaNosP* from *Pseudomonas aeruginosa* with independent protein units functioning as an NO sensor and the responsive unit. The different general modes of operation of metal-based gas sensors are illustrated, where the first case involves catalysis using O_2 as shown by the O_2 sensor HIF hydroxylase. The next mode employs iron–sulfur cluster site disassembly represented by the O_2 sensor FNR, while the last case involves reversible binding to the gaseous signaling molecule as illustrated by the O_2 sensor DosP.

There are also many types of metallosensing units based on how they respond to the biological signal, as described elsewhere [1]. However, most of them function as a catalytic unit, a metal disassembling unit or a reversible binding unit [1] (Figure 1). The O_2 sensor HIF hydroxylase from *Homo sapiens* functions by catalyzing O_2-based hydroxylation of HIF-1α (a transcription factor), leading to its degradation, turning off a genetic hypoxia response [12,13]. Iron–sulfur cluster unit found in the O_2 sensor FNR from *E. coli* function by reacting with O_2 disassembly of the iron site [9,14]. However, many types of metalloprotein gas sensors function upon reversible and direct binding of the gaseous signal to the metal site such as O_2 sensor FixL that employs a heme cofactor for reversible binding [3]. Besides all of that, a validated sensing unit upon binding to these gaseous molecules must report such process to a responsive unit (also known as a functional output structure), leading to biological adaptation.

The responsive unit is another essential part of these gas sensors, being involved in the conversion of the gas binding process into a biochemical signal/information. Similarly to the sensing unit, it can also be found as part of a certain region of the protein, in one different domain or in another independent protein (Figure 1). There are many functional output units used by nature: those with enzymatic activity (e.g., histidine kinase [10,15,16], nucleotide cyclase [17–19], phosphodiesterase [18,19]), capable of interacting with a protein or DNA and others [20], meaning that the binding of O_2, NO and CO can regulate the production or degradation of molecules (e.g., c-di-GMP by DosC/DosP impacting RNA [21,22]) or the interaction with a DNA sequence enabling gene regulation (e.g., CooA [23]) or modification of a protein for degradation (e.g., HIF hydroxylase [24]), among other things [1,2,20]. These intricate arrangements have allowed nature to come up with many outstanding biological responses, regulating multiple processes as important as blood pressure, biological clock, nitrogen fixation, dormancy, bacterial biofilm, RNA degradation, virulence, among others. Unfortunately, a detailed structural mechanism of the signal transduction processes is still mainly lacking, but some important insights have emerged, as mentioned briefly here and in detail elsewhere [1,2,20,25–29].

The largest family of metalloprotein gas sensors is based on heme-containing proteins [4]. These heme-based gas sensor proteins have been commonly found containing multi-domains (e.g., FixL, DosP), but there are also standalone proteins as well (e.g., *Pa*NosP, Figure 1) [30]. Soluble guanylate cyclase (sGC) was the first discovered case of a heme-based gas sensor protein, which functions as a dedicated nitric oxide (NO) sensor [17,31]. This protein contains a heme domain, one intervening domain and an enzymatic domain that converts GTP into the second messenger cGMP and pyrophosphate (PP_i). This enzymatic activity is strongly enhanced upon NO binding to the heme domain, leading to major production of cGMP [17]. This latter molecule is a secondary messenger involved in the regulation of many physiological processes [32]. Actually, sGC is involved in many important processes in humans and other organisms, including vasodilation, platelet aggregation, memory, etc. [33,34].

In 1991, Gilles-Gonzalez discovered another heme-based gas sensor called FixL [10]. This protein was shown to be an O_2 sensor. FixL is an important upstream oxygen sensor for *Rhizobia* bacteria, which allows symbiosis to occur with leguminous plants [35]. That important discovery was soon followed by others, supporting the existence of a superfamily of heme-based gas sensors. Currently, there are at least eight major families of these sensors based on the heme fold domain, which includes the PAS, HNOB (or HNOX), globin, CooA (CRP), SCHIC, GAF, LBD (Holi) and FIST domains (Figure 2) [36,37]. This latter family is the most recent one discovered in 2017 [11], whose sensors are called NosP (nitric oxide-sensing protein). The most common diatomic gas type sensed by each family is shown in Figure 2, along with some well-known protein sensors and their biological functions.

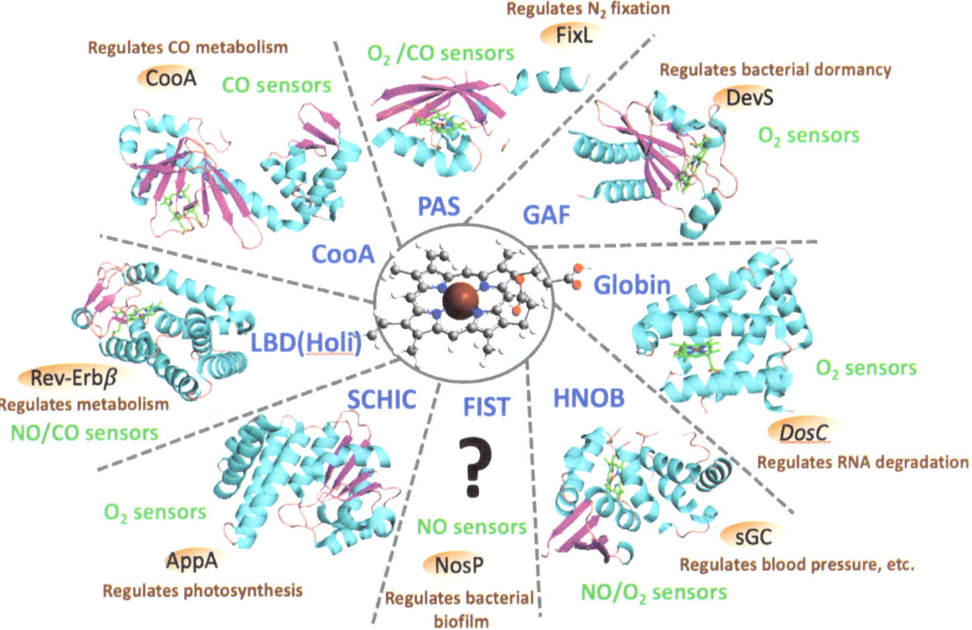

Figure 2. Heme-based gas sensor families, structural heme fold (in blue; PAS domain (pdb 1DP9), GAF domain (pdb 4YNR), globin domain (pdb 6OTD), HNOB domain (pdb 2O09), FIST domain (3D structure not solved yet), SCHIC domain (pdb 4HEH), LDB(Holi) domain (pdb 3CQV), CooA domain (pdb 2FMY)), common functionality as a gas sensor (green) and some examples of proteins of each family (brown oval) and biological function (brown).

2. FixL—A Short Story

Nitrogen fixation promoted by some bacteria is a very important process that allows proper access of a nitrogen source to plants during a symbiotic process [35,38]. This process is usually complex, and nitrogenases are also oxygen-sensitive, requiring many physiological changes to happen within the plant nodule and *Rhizobia* [38]. The plant is responsible for building an oxygen barrier in the nodule along with a large production of leghemoglobin, creating extremely low levels of free oxygen (nanomolar) [39]. In this way, it makes it possible for a nitrogenase to work. The level of oxygen within the nodule is closely monitored by FixL, which under virtual anaerobiosis becomes enzymatically active [35]. Thus, FixL can phosphorylate FixJ, another protein that functions as a transcription factor regulating genes involved in microaerobic respiration and the nitrogenase apparatus [35]. This upstream oxygen sensor along with NifA assure that a cascade of events takes place, leading to nitrogen fixation [35,38].

FixL is a multidomain sensor protein that contains a heme b group in a PAS fold domain and a histidine kinase domain in its C-terminal region (Figure 3A). Sometimes, this sensor exhibits an extra non-heme PAS domain or transmembrane segments at the N-terminal region [5]. Despite the lack of an X-ray structure for this full-length protein, there are many structural and functional studies on this sensor that have elucidated important molecular details [40–54]. One interesting feature is the ability of this protein to recognize O_2 and selectively trigger a response to the histidine kinase domain. O_2 as well as CO and NO binds to the iron heme, where O_2 has the weakest affinity of all, but at the same time it is the only one that switches off the protein enzymatic activity [15,45]. This is associated to its geometry upon binding and specific interactions with distal residues [41,53,55]. This subtle binding of O_2 causes conformational changes within the heme domain, altering the heme distortion (induced planarity) and leading to specific interactions with arginine and heme propionate groups within the PAS heme domain [41,47,53]. These changes have been associated to the signal transduction process that propagates to the enzymatic domain. Recently, SAXS studies suggested signal transduction occurs with a more subtle movement of the domains [40], not necessarily by switching contacts. A movement of the coiled-coil linker region that connects the heme PAS and histidine kinase seems to be vital for signal transduction, affecting the proximity of the ATP binding site and the histidine phosphorylatable residue [40].

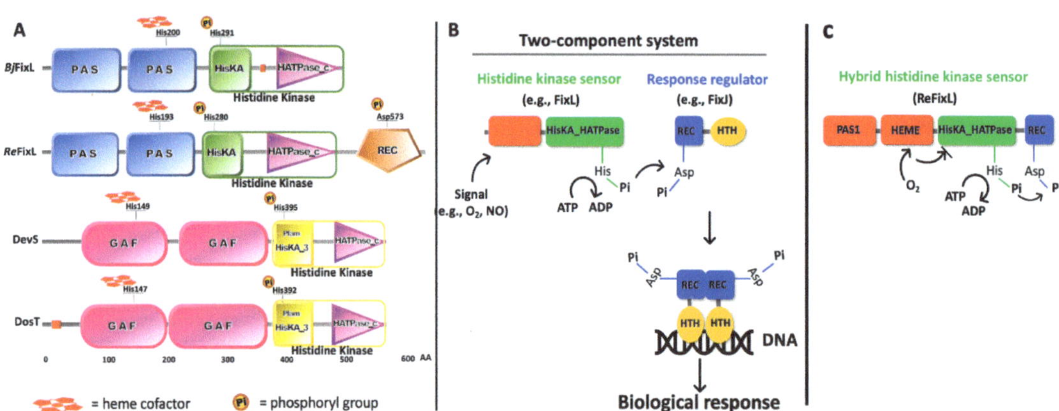

Figure 3. (**A**) Domain organization of FixL and DevS/DosT (heme as a red small structure, phosphorylation indicated by Pi; heme proximal ligand and phosphorylatable residues are shown) composed of the PAS and GAF domains along with the histidine kinase domain constituted of the HisKA and HATase_c subdomains. Hybrid FixL (*Re*FixL) presents an additional REC domain (receiver domain) usually found at the N-terminal region of the response regulator proteins. (**B**) Cartoon of the functioning of common two-component systems, where a histidine kinase sensor upon the appropriated signal becomes active, autophosphorylating a histidine residue; then, a phosphoryl group is transferred to the response regulator protein. This later becomes active upon phosphorylation and usually functions as a transcriptional factor triggering a biological response. (**C**) Autophosphorylation of the hybrid *Re*FixL, where O_2 blocks histidine kinase and phosphotransferase activities. His = histidine residue, Asp = aspartate residue, AA = amino acids.

The bacteria have commonly used a particular type of sensing systems to perceive and respond to environmental changes called two-component systems, which are usually formed by one histidine kinase sensor and one response regulator protein (Figure 3B) [56]. FixL and FixJ are a typical two-component system responsible for sensing O_2, where FixL is a histidine kinase sensor and FixJ is a response regulator. How these multi-domains communicate and coordinate intra- and interprotein phosphorylation are still exciting questions. The presence of additional "nonfunctional" domains in some FixL proteins raised further question on their structural roles. FixL from *Bradyrhizobium japonicum* (*Bj*FixL) is such a case, where one additional non-heme PAS domain is found at the N-terminal region followed by a heme-PAS domain and a histidine kinase (Figure 3A). The actual role of this additional PAS domain is not clear, but it may be quite important as shown in another study using a hybrid FixL sensor from *Rhizobium etli* (*Re*FixL) [57]. This latter cytosolic sensor exhibits a domain organization quite similar to *Bj*FixL (Figure 3A), but it also has an extra domain at the C-terminal region. This extra terminal domain (REC) is analogous to the receiver domain of FixJ, which contains a phosphorylatable aspartate residue (Figure 3A). Thus, this protein features as a combination of *Bj*FixL and part of *Bj*FixJ, being a great opportunity for questioning the function of multi-domains in the FixL–FixJ system. In such a hybrid FixL sensor, there are two phosphorylatable residues, one histidine residue centered at the kinase domain and one aspartate at the C-terminal receiver domain (Figure 3C). Then, the hybrid protein can promote histidine autophosphorylation followed by an intramolecular phosphotransfer to aspartate in the receiver domain, all within the same protein.

This hybrid *Re*FixL sensor brought up an opportunity to explore this cascade of phosphorylation processes and their regulation [57]. Interestingly, this sensor showed the lowest reported oxygen affinity for a heme-based gas sensor ever ($K_d = 738$ μM) [57], despite *Bj*FixL also having a reasonably low affinity, too ($K_d = 140$ μM) [58]. Despite that, *Re*FixL exhibited very tight oxygen regulation even in the air, where only 26% of the protein is bound to oxygen. The deletion of the first PAS domain (ΔPAS1 *Re*FixL) provoked major changes not only in the heme domain, but also in the histidine kinase activity. While the oxygen affinity increased about eightfold, the histidine kinase activity was completely turned off with minor measurable ligand regulation [57]. This lack of activity could be due to a damaged inactive histidine kinase domain; however, ΔPAS1 *Re*FixL was still capable of being phosphorylated by another fully functional ReFixL. This result supported the histidine kinase domain was still functional, but ΔPAS1 ReFixL was likely shifted toward an inactive equilibrium state. The PAS1 domain seems to allow an equilibrium between active and inactive states to be achieved, whose removal disrupted this conformational change, locking the protein in an inactive state. In another study, we also noticed the plasticity of their domains and differences in the overall stability when measured in the active versus inactive state [48]. These studies illustrated how sensitive the heme domain is and how major properties can be altered by interaction with other domains, not only by mutations within the heme domain itself. Indeed, this particular feature has been noticed by many other studies comparing heme properties of isolated heme domains with full-length proteins [1], reinforcing sensing domains are much more flexible as they should be to promote signal transduction.

Another exciting feature discovered in FixL is an O_2 memory effect. This phenomenon is a hysteresis response caused upon O_2 binding, which quickly inactivates histidine kinase, but FixL reactivation takes much longer once O_2 dissociates [59]. There was an initial expectation that one O_2 molecule bound to FixL is capable of turning off its histidine kinase activity. If this behavior was right, a linear dose response would be obtained for O_2 saturation of FixL versus histidine kinase activity. However, experimental data showed the dose response for *Bj*FixL was remarkably nonlinear, but it matched nicely the in vivo O_2 dose response regulation data. In other words, this result means that a few molecules of O_2 are able to inactivate many molecules of FixL due to faster O_2 rebinding in comparison to slower kinetics of histidine kinase activation. This process fits well in a hysteresis model supporting in vitro and in vivo data. This exciting phenomenon can indeed explain how extremely low-affinity sensors work [57,59]. Actually, the nonlinear response of inhibitors (as drugs) has alerted medicinal chemists of an important aspect of association/dissociation kinetics beyond thermodynamic affinity (K_d) [60]. The memory effect may explain how oxygen can regulate *Re*FixL so tightly despite its extremely low affinity. While this process may explain FixL behavior, it was not able to explain the nonlinear dose response for other oxygen sensors (e.g., DosP [19]), where other mechanisms may take place.

Structural changes of the heme domain of FixL were observed as reasonably fast (few microseconds) events associated with planarity/distortion of the heme and its interactions with nearby residues [61–63]. Recently, an attempt to investigate the time scale of the signal transduction process was carried out with FixL from *Sinorhizobium meliloti* (*Sm*FixL) using Raman spectroscopy [64]. Despite the limitations of this study, the authors suggested conformational changes provoked in the heme domain by oxygen binding would take at least microseconds to alter the histidine kinase domain. However, it is still unknown if any further fine adjustment is required to achieve full inactivation or activation.

FixL or analogs have been found in other microorganisms, such as *Caulobacter crescentus* [65], *Brucella abortus* [66], *Chlamydomonas reinhardtii* [67] and *Burkholderia cepacia* [68,69]. This latter bacterium is part of a class of pathogens causing concerns in cystic fibrosis patients. There is growing evidence that this bacterium employs the oxygen sensor FixL to control biofilm formation, motility, intracellular invasion/persistence and virulence, opening exciting opportunities to explore this system in therapy. Beyond that, this system has also been explored in synthetic biology as mentioned later (see section "Diverse Potential Applications of Heme-Based Gas Sensors: Systems Applied in Cell Biology").

3. DevS and DosT—A Short History

The millenary disease, tuberculosis, is still a major global health issue, causing deaths of over 1 million people around the world [70]. The low number of drugs, emergence of multiple drug-resistant strains and the large reservoir of dormant bacteria within humans have caused a challenge to control and eradicate this disease [71,72]. Almost 20 years ago, a two(three)-component system (DevS–DevR, and later DosT) was identified in *M. tuberculosis* (Mtb) [73–75], which was associated to its virulence. Genetic studies showed this system was oxygen-responsive and associated to the initiation of the dormancy program in the bacterium [74,76]. The dormancy of Mtb is related to its persistence in humans, the lengthy medical treatment and difficulty to eradicate this disease in the world [77]. Importantly, this sensory system of DevS–DosT–DevR was shown as a suitable drug target, where two independent laboratories have been developing new compounds as novel anti-tuberculosis drugs (see section "Diverse Potential Applications of Heme-Based Gas Sensors: Pharmacological Use").

In 2005, DevS (also known as DosS) was shown to be a GAF heme-containing protein [78] (Figure 3B). Later, DevS and its ortholog, DosT, were demonstrated to function as O_2 sensors [79,80], supporting previous genetic studies [74,76]. In 2007, another laboratory suggested DevS is a redox sensor while DosT is a direct oxygen sensor [81]. These apparently conflicting conclusions were addressed by other studies as discussed elsewhere [36]. In summary, there is strong and compelling evidence supporting that DevS and DosT are indeed direct oxygen sensors, including DevS having a relatively high electrochemical potential ($E_m = -10$ mV vs. NHE) [82]. But why nature would employ two oxygen sensors for that process? These proteins may function by sensing and responding to distinct levels of hypoxia, which would be in agreement with their oxygen affinity [79], as indicated by genetic studies [83]. DosT with a lower oxygen affinity ($K_d = 26$ μM) may perceive the initial drop in oxygen concentration, allowing initial bacterial adaptation. If this condition does not persist, it may be an easier return to normal growth; however, if a further decrease in the oxygen level takes place, then DevS becomes active, leading to major physiological changes into a dormant state. Despite their differences, the accrued biochemical knowledge on FixL–FixJ helped us to better understand this system [1,3,4].

Raman spectroscopy, site-direct mutagenesis and single crystal X-ray studies provided important hints on the role of key residues involved in sensing and signal transduction [84–86]. The distal ligand in the heme domain is a tyrosine residue (Y171 in DevS and Y169 in DosT), which makes a hydrogen bond with O_2, supposedly assisting its binding. It is interesting to remark that such assistance was not validated for DevS, where the Y171F mutant showed an even higher affinity to oxygen [85], illustrating the complexity of these interactions. Unlike FixL, these proteins showed minor changes within the heme pocket in the active and inactive states, besides a modification in the hydrogen bonding network from the heme to the surface of the domain [86]. This effect might report O_2 binding to the surface of the domain and could propagate this information into other domains as a signal transduction process. Other studies were carried out to identify the role of other residues within the heme, where arginine 204 seemed to have a critical role [87]. This residue is expected to be in a polar patch at the surface of the heme domain in an interface with the other GAF domain. Notably, the R204A mutant promoted a complete disruption in the histidine kinase activity, implicating this residue in signal transduction [87]. Earlier, another study had investigated the role of the extra GAF domain, where two constructs were compared with the wild-type one containing the two GAF domains (GAF(A)–GAF(B)) and another one with only the heme GAF domain (GAF(A)) [88]. These constructs in comparison to wild-type DevS showed distinct features by Raman spectroscopy. Interestingly, the heme GAF domain upon binding to CO exhibited a Raman spectrum consistent with two conformations, while the two GAF constructs had only one, indicating that the intervening domain might mediate signal transduction initiated in the heme domain.

Recently, our laboratory showed DevS is a mixture of oligomers dimer–tetramer–octamers, and more importantly, this oligomeric distribution could be shifted by selective disturbance in the heme domain [89]. Indeed, active DevS either under anaerobic (deoxy-DevS) or CO-bound form (CO-DevS) exhibited a major fraction as octamers, while in the inactive state, either ferric (met-DevS) or oxygen-bound forms (oxy-DevS) showed mainly tetramers and dimers. Remarkably, the inhibition ratio (histidine kinase activity in the active/inactive states) matched nicely the relative amount of octameric/(dimeric–tetrameric) species, supporting the phenomenon of signal transduction promotes changes in the quaternary state of the protein. This process was also reasonably fast (a few minutes), and may occur in other heme-based gas sensors as well [90,91].

The overall mechanism of functioning of this two(three)–component system may be more complex than expected [25]. There is a series of other protein partners, protein modifiers (serine threonine kinase, acetylation) and cross-talks with other sensing systems involved, making this whole story complex, as discussed elsewhere [25]. Nevertheless, there are many opportunities ahead to unfold these details and find new applications for these systems either as drug targets or biological tools, etc.

4. Diverse Potential Applications of Heme-Based Gas Sensors

Our understanding of the molecular functioning of heme-based gas sensors and their broad diversity has also opened new opportunities, both to develop small-molecule regulators and use them as new tools and materials, as we discuss further (Figure 4).

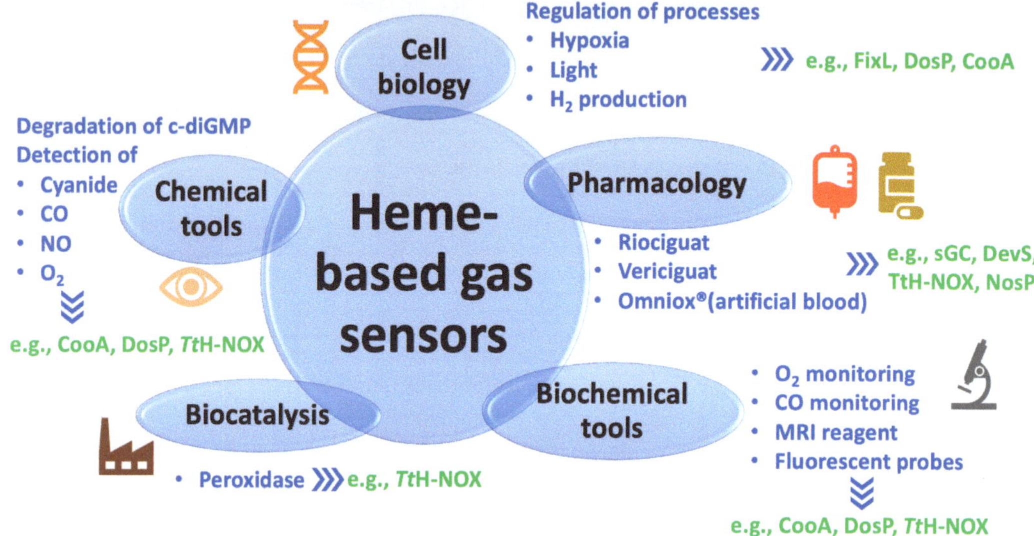

Figure 4. Chemical and biotechnological applications of heme-based gas sensors. There are five segments of investigations with some developed applications of heme-based gas sensors in cell biology (e.g., FixL was employed to achieve hypoxia and light control), pharmacology (e.g., two FDA-approved drugs are available targeting sGC, while others are under study, along with a blood substitute candidate Omniox®), biochemical tools (e.g., CooA, DosP and *Tt*H-NOX are explored for in vivo monitoring of O_2, CO and other uses such as MRI or fluorescent reagents), biocatalysis (e.g., *Tt*H-NOX has been explored as a peroxidase catalyst), chemical tools (e.g., CooA and *Tt*-HNOX have been explored for in vitro detection and measurement of cyanide, CO, NO and O_2).

4.1. Pharmacological Use

Currently, there are many heme-based gas sensors involved in key biological processes both in humans and pathogens, making them suitable drug targets for investigation [92]. Here, we described a few interesting cases, some of them still in very early stage while others with drugs approved for clinical use.

The first heme-based gas sensor discovered, sGC, has many physiological roles in humans, making it a very desirable drug target for the treatment of endothelial, cardiovascular and pulmonary disorders [93,94]. A series of nitric oxide (NO) donors was earlier employed targeting sGC, even before this sensor was known or this small molecule (NO) found to be biologically active. The binding of NO to sGC strongly stimulates its cyclase activity, converting GTP into cGMP [27]. This is one of the reasons for the development of NO donor molecules, which can be employed to lower blood pressure, hypertension crisis, angina pectoris, glaucoma, among other applications [94]. However, the lack of selectivity and acquired tolerance to NO can limit its action in continuous use. Some strategies have emerged to deal with these issues, particularly providing more selective delivery strategies either by using light, nanoparticles or other stimuli [1,94]. At the same time, other small molecules have been investigated, starting with YC-1, the first direct sGC stimulator discovered [95]. Currently, there are many organic- and inorganic-based molecules investigated as sGC stimulators or activators likely able to treat many conditions [92–94,96]. Two drugs (riociguat, 2013, and vericiguat, 2021) that function by targeting sGC are already approved for clinical use, opening many opportunities. Other heme-based gas sensors found in humans have also been explored as potential drug targets involved in the mood and metabolic disorders (e.g., NPAS2, Rev-erbα/β) [92,97].

The disruption of sensing systems in bacteria has emerged as a promising strategy for the development of novel antibiotics with likely minor selective pressure [98]. Among these systems, there are heme-based gas sensor proteins as potential drug targets [1,92]. In certain cases, the use of NO donors can be beneficial to assist with the elimination of the bacterial infection through interaction with heme-based gas sensors [94]. One of the major obstacles to eliminate bacteria is caused by the production of biofilm that leads to antibiotic tolerance [99]. The NO sensor NosP found in *P. aeruginosa* and *V. cholerae* is involved in the regulation of biofilm production, where binding to NO promotes biofilm dispersal [100]. This has opened new opportunities for the use of NO donors, particularly in combination with antibiotics. Recently, we and others have been able to show that NO donors can function in synergy with antibiotics, promoting the disruption of biofilms and enhancement of antibiotic action [101–103].

The O_2 sensor two(three)–component system of *M. tuberculosis*, DevS/DosT–DevR, has also become an important drug target [72,92,104]. Two laboratories have made efforts to identify new molecules to inhibit this sensing system. Tyagi's laboratory pioneered this study in identifying phenylcoumarin-based compounds through an initial in silico screening (2.5 million compounds), but their hits exhibited low biological activity [105]. Then, they carried out other studies looking at small peptides targeting DevR and DevS using the phage display technology [106,107]. Unfortunately, the most active peptides identified did not show suitable biological efficiency. Recently, Abramovitch's laboratory developed a cell-based phenotypic screening assay with a reporter gene under DevR regulation [108]. In this study, they employed a library of 540,288 compounds, where six classes of compounds were identified with apparent distinct mechanisms of action. These compounds showed quite promising biological activity toward Mtb with apparently distinct mechanisms of action, including a synergistic effect with anti-tuberculosis drugs (e.g., isoniazid) [104,108,109]. One interesting case was with artemisinin, one class of drug found in the screening, that was shown to target the heme of DevS and DosT [108]. Other laboratories had shown the potential anti-tuberculosis activity of this old compound also in combination with isoniazid (and other anti-TB drugs) [110,111], but there was no clue about its potential target. These exciting studies may continue and eventually provide a leading compound targeting dormant bacteria, a major issue during tuberculosis treatment. Besides this case, other heme-based gas sensors might function as a drug target in the development of antibiotics or antiparasitic agents, deserving a close look, including in certain FixL systems (found in the pathogens *Burkholderia cepacia* and *Brucella abortus*) [66,68]), *Bpe*GReg (found in *Bordetella pertussis*, a pathogen that causes whooping cough) [112], Hpk2 (found in *Treponema denticola* that causes periodontal disease) [113], DosC/DosP (found in *E. coli*) [19,89] and HemAC-Lm (found in the parasite *Leishmania*) [114].

Beyond this, due to the large diversity of heme domains reported for heme-based gas sensors, they should be further investigated as potential antidotes for carbon monoxide poisoning, delivery of gasotransmitters (e.g., Sanguinate® [115]) or even as blood substitutes as carried out with neuroglobins [116] and modified hemoglobins (e.g., Hemopure®) [117]. There is one expired patent on the potential use of globin-based proteins including heme-based gas sensor proteins HemAT from *Halobacterium salinarum* and *Bacillus subtilis* in medicine as blood substitutes and microsensors [118]. As far as we are aware, there is only one case of a heme-based gas sensor protein being investigated as a blood substitute, which was reported for the H-NOX sensor from *Thermoanaerobacter tengcongensis* (*Tt*H-NOX, Omniox®) [119]. In this study, a modified *Tt*H-NOX heme domain exhibited promising results in preserving myocardial contractility during acute hypoxia in an ovine model of myocardial injury, whereas NO sequestration was not an issue, commonly faced when using hemoglobin-derived biotherapeutics. This heme-based gas sensor has been explored in many other applications as described further.

4.2. Chemical and Biological Tools

DosP (earlier called *Ec*DOS) is a heme-based O_2 sensor from *E. coli*, which upon binding to oxygen exhibits a steep increase in c-di-GMP phosphodiesterase activity [120]. This dinucleotide is an important signaling molecule in bacteria with important roles in humans as well [121]. A recent study showed that DosP can be efficiently encapsulated into mesoporous silica, where its regulated phosphodiesterase activity is reasonably preserved along with stability [120]. This type of material might have interesting chemical and biological applications. The easier chemical production of linear pGpG from c-di-GMP phosphodiesterase might be one potential use of this material. In addition to that, its capacity to degrade c-di-GMP might have therapeutic features against microbes that could be explored [122].

The analytical use of heme-based gas sensors as probes for diatomic gases (O_2, NO and CO) or anions (cyanide) has been explored. The heme-based CO sensor from *Rhodospirillum rubrum*, CooA, was designed to report CO levels in vitro and in vivo [123]. This was achieved by fusing a fluorescent protein in the C-terminal domain, where a major conformational change occurs when CO binds. Notably, this sensor exhibited an increase in fluorescence upon binding to CO with a K_d of 2 µM, which was minimally disturbed by NO, O_2, H_2S, imidazole, glutathione or CN^-. This probe was also investigated in vivo by transfecting HeLa cells with a probe-expressing vector, where fluorescence images were obtained either using 5 µM CO (externally applied) or a known CO donor molecule (CORM-2, used even at 1 µM).

A mutant of the H-NOX sensor from *Thermoanaerobacter tengcongensis* (*Tt*H-NOX Y140F) was developed to function as an NO sensor. This mutant exhibited great properties such as no binding to oxygen, high stability to heat (up to 70 °C) and oxidation, but a very strong affinity to NO. Notably, this protein can be used to measure very low levels of NO even in a large background of O_2. Another modification of *Tt*H-NOX was carried out by replacing its native heme with a ruthenium(II)-mesoporphyrin IX [124]. In this case, the protein scaffold hosted a luminescent ruthenium-based cofactor that upon binding to O_2 exhibits light emission suppression, functioning as an oxygen sensor. Indeed, the authors showed they could detect up to 4.2 µM O_2, claiming this is reasonably close to commercial O_2 sensors [124].

In another study, meso-alkynylation of heme b cofactor was carried out, allowing further conjugation with many molecules, including fluorophores. The authors showed this modified porphyrin could be efficiently incorporated in the *Tt*-H-NOX protein, then labeled with an azide-based fluorophore (rhodamine) [125]. This procedure offers more flexibility where labeling is not in the protein chain, making it possible to explore fluorescence imaging as well as Raman microscopy. In a recent study, a highly fluorescent near-infrared compound, phosphorus corrole, was investigated as replacement of the native heme of the *Cs*H-NOX sensor from *Caldanaerobacter subterraneus* [126] (formerly known as *Thermoanaerobacter tengcongensis* H-NOX or *Tt*-H-NOX sensor [127]). The authors showed phosphorus corrole was able to replace its native heme cofactor, providing better properties such as long wavelength emission and high quantum yield. Corrole-based compounds exhibit some structural similarity with protoporphyrin IX, but they can be much better emitters, which in combination with a robust protein scaffold can generate promising new fluorescent systems [126]. Besides these cases, this same heme-based gas sensor was also explored in other applications, such as cyanide sensor (*Tt*H-NOX P115A mutant) [128] and MRI probes using Mn(II/III) and Gd(III) porphyrins as a substitute to the native heme [129].

4.3. Systems Applied in Cell Biology

An early study produced a sensory chimera (CskA) containing part of FixL [130], aiming mostly to understand the essential elements for building a functional sensor. These investigators combined the oxygen-sensing heme domain of *Sm*FixL from *Sinorhizobium meliloti* with the histidine kinase domain of ThkA from the hyperthermophile *Thermotoga maritima*. In that work, two out of five constructs showed oxygen affinity close to full-length *Sm*FixL, but none of them exhibited oxygen regulation of histidine kinase. Nonetheless, some constructs showed even higher autophosphorylation activity if compared to ThkA [130], indicating modulation of the enzymatic activity in the chimera. This frustrated attempt to prepare an oxygen-regulated system illustrated the key role of the linkers to allow proper interaction of the domains and likely allow regulation. Other attempts were more successful in achieving signal regulation as described below.

In 2008, Moffat's laboratory developed a chimeric protein by replacing the heme domain of the oxygen sensor FixL with the light–oxygen–voltage-sensing domain (LOV domain of YtvA from *Bacillus subtilis*) [131]. Their goal was to control reversibly the histidine kinase activity of FixL using blue light instead of oxygen. Interestingly, they prepared some protein constructs with remarkable regulation of enzymatic activity using blue light, achieving over 1000-fold inhibition. This system was also investigated in *E. coli*, where light irradiation was able to shut off 70-fold a gene reported based on this system [131]. This study opened exciting opportunities for a broad use of sensing systems. Another independent laboratory prepared a chimera of FixL with CckA, this latter being a histidine kinase sensor from *Caulobacter crescentus*, as a biological tool [132]. This system combined the sensing unit of CckA with the histidine kinase of FixL and full-length FixJ along with a fluorescent gene reporter responsive to phosphorylated FixJ. The use of this biological tool allowed them to better understand the molecular signaling mechanism associated with the asymmetric cell division of *C. crescentus* [132].

Another chimeric protein was developed using the heme PAS domain of DosP, an oxygen sensor in *E. coli*, combined with a yellow fluorescent protein (Venus YFP) [133]. This new protein, called ANA-Y (anaerobic/aerobic sensing yellow fluorescence protein), has a coiled-coil linker optimized to provide suitable fluorescence response upon oxygen binding to the heme domain. This system can report oxygen concentrations in the micromolar range, which was employed as a genetic probe in synthetic biology to monitor initial photosynthetic production of oxygen in cyanobacteria [133]. In the previous subsection (see "Diverse Potential Applications of Heme-Based Gas Sensors: Chemical and Biological Tools"), we also presented a CO probe based on a chimeric protein of CooA. This sensor regulates bacterial processes involving the consumption of CO with the generation of H_2 such as in the water–gas shift reaction ($CO + H_2O \rightarrow CO_2 + H_2$). This process is catalyzed by two enzymes, carbon monoxide dehydrogenase (CODH) and carbon monoxide-dependent hydrogenase (CO-Hyd) under regulation by CooA [134]. Aiming to enhance the production of H_2 by *Citrobacter amalonaticus* Y19, an engineered strain was prepared containing a plasmid carrying the cooA gene for overexpression [134]. This genetic alteration promoted an expressive increase in H_2 production (3.4-fold) by this bacterium, supporting this type of biotechnological strategy.

4.4. Biocatalysts

In 2018, Dr Frances Arnold was awarded a Nobel Prize in Chemistry for her contribution to the development of novel biocatalysts based on directed evolution [135]. Despite that, biocatalysts have been developed not only based on directed evolution (random mutations), but also with exchange of metal ions and incorporation of new cofactors that have broadened our horizon. Indeed, enzymatic catalysis of reactions not reported to be carried out by nature were developed by designing novel biocatalysts. Myoglobin, a non-enzymatic hemeprotein, has been widely explored and converted into an outstanding novel catalyst for many distinct biological and nonbiological reactions, as reviewed elsewhere [136]. The same has been carried out for many other metalloproteins, including other hemeproteins (e.g., cytochrome c, cytochrome P450). This may be a great opportunity to explore many distinct heme domains found in the families of heme-based gas sensors. As far as we are aware, there is only one case explored so far, as described below.

Aiming to explore and enhance potential catalytic properties of the heme-based gas sensor *Tt*-H-NOX, its heme domain was studied and subjected to site-direct mutagenesis at the distal tyrosine ligand [137]. Interestingly, this protein showed capacity to decompose H_2O_2, which was strongly enhanced in the Y140H mutant and lowered in the Y140A mutant. Their peroxidase activities were investigated using the ABTS (2,2′-azino-bis(3-ethylbenzothiazoline-6-sulphonic acid)) substrate along with H_2O_2, where the WT protein showed moderate activity but the Y140H mutant was significantly better (70% higher) in degrading ABTS. Despite this protein showed a much lower turnover compared to other known catalytic proteins, the authors claimed there is a comparable efficiency in oxidizing substrates as in some native catalysts (e.g., but using different substrates, *Tt*H-NOX $k_{cat} = 0.06 \ s^{-1}$ for ABTS versus cytochrome P450BM-3 $k_{cat} < 0.008 \ s^{-1}$) [137]. This study did not provide appealing catalytical results, but it highlighted the opportunities ahead, taking advantage of the thermostability, easy production and structural robustness of that heme-based gas sensor protein.

In another study, a multidomain protein containing one heme PAS domain showed an unexpected oxidative N-demethylase activity based on the heme cofactor [138]. The PAS domain is indeed a typical signal transduction unit employed by many types of sensors, including heme-based gas sensors, which has never exhibited any reported catalytic activity. Despite this protein might not be a true heme-based gas sensor, it has exposed the heme PAS domain as a biocatalytic unit deserving further studies. This unique case highlights the broad opportunities to explore biocatalysis using many other types of heme domains from heme-based gas sensors.

As shown here, there are many opportunities to create novel tools and materials by manipulating heme domains. One strategy to explore this is to replace the native heme by another metalloporphyrin analogs, providing novel functions to the protein as described before. To achieve this, engineered *E. coli* strains or growth conditions have been explored to facilitate direct incorporation of heme analogs or production of hemeless proteins [139–142]. The engineered RP523 strain of *E. coli* was prepared, exhibiting disruption of biosynthesis of the heme (hemB and hemC) along with the heme permeability phenotype [139]. This feature allowed direct incorporation of metalloporphyrins once they were provided in the cell culture medium [139], but oxygen sensitivity and the toxicity of cofactors are a concern for its success. Another strategy was to use an *E. coli* BL21(DE3) strain containing a plasmid bearing the heme transporter ChuA allowing the uptake of heme analogs provided in the cell medium [141]. In addition to that, *E. coli* was grown in an iron-depleted medium to minimize heme biosynthesis, but up to 5% of the protein could still bear the heme. Alternatively, it was shown that the hemeproteins overexpressed in *E. coli* BL21(DE3) employing the M9 medium are mainly hemeless, where the incorporation of exogenous metalloporphyrins can be easily achieved by mixing with lysed cells [140]. These approaches are important technical advances helping the success of using heme-based proteins.

5. Final Considerations

Heme-based gas sensor proteins are a remarkable widespread superfamily of sensing systems. These moderately old sensors always surprise us with their diversity, appearance of new families and mechanisms of action. There is no doubt many more are still going to be discovered. The multiple applications of these systems are just unfolding, covering biological and chemical tools, biocatalysis and drug targets, making them a hot spot of opportunities much beyond our necessity of understanding their molecular functioning.

Author Contributions: Conceptualization E.H.S.S.; Writing—original draft preparation: E.H.S.S.; Writing—review and editing: A.C.S.G., W.G.G. and E.H.S.S.; Project administration: E.H.S.S.; Supervision E.H.S.S.; Visualization: A.C.S.G., W.G.G. and E.H.S.S. All authors have read and agreed to the published version of the manuscript.

Funding: This work was funded by CNPq (EHSS 308383/2018-4), the National Institute of Science and Technology on Tuberculosis (INCT-TB, financed by CNPq/FAPERGS/CAPES/BNDES) and Coordenação de Aperfeiçoamento de Pessoal de Nível Superior-Brasil (CAPES), Finance Code 001 (PROEX 23038.000509/2020-82).

Institutional Review Board Statement: Not applicable.

Informed Consent Statement: Not applicable.

Acknowledgments: We wish to thank all the organizers of XXI SPB National Congress of Biochemistry 2021 that took place in Évora (Portugal) for the high-quality meeting and the opportunity to deliver a talk on heme-based gas sensors and also contribute to this special issue. We are also thankful to CNPq (EHSS 308383/2018-4) and the National Institute of Science and Technology on Tuberculosis (INCT-TB, financed by CNPq/FAPERGS/CAPES/BNDES) for financial support. This study was also financed in part by Coordenação de Aperfeiçoamento de Pessoal de Nível Superior-Brasil (CAPES), Finance Code 001 (PROEX 23038.000509/2020-82).

Conflicts of Interest: The authors declare no conflict of interest.

Abbreviations

*Bpe*GReg	*Bordetella pertussis* globin-coupled regulator
CckA	*Caulobacter crescentus* histidine kinase protein A
c-di-GMP	Cyclic dimeric ($3',5'$) guanosine monophosphate
cGMP	Cyclic guanosine monophosphate
*Cs*H-NOX	H-NOX sensor from *Caldanaerobacter subterraneus*, formerly known as *Tt*H-NOX
CskA	Chimeric sensory kinase A
CooA	CO oxidation activator protein
CORM-2	CO-releasing molecule 2
DevS	Differentially expressed in virulent strain sensor protein, the same protein as DosS
DevR	Differentially expressed in virulent strain response regulator protein, the same protein as DosR
DosC	Direct oxygen sensor cyclase protein
DosP	Direct oxygen sensor phosphodiesterase protein
DosS	Dormancy survival sensor protein
DosT	Dormancy survival sensor T protein
*Ec*DOS	*Escherichia coli* direct oxygen sensor protein, the same protein as DosP
FIST	F-box and intracellular signal transduction proteins domain
FixL	Rhizobial nitrogen fixation gene L protein
FixJ	Nitrogen fixation gene J protein
FNR	Fumarate and nitrate reduction protein
GAF	cGMP-specific phosphodiesterases, adenylyl cyclase, and FhlA proteins domain

GTP	Guanosine-5′-triphosphate
HemAC-Lm	Heme-containing adenylate cyclase from *Leishmania major*
HemAT	Heme-based aerotactic transducer sensor protein
HIF	Hypoxia-inducible factor
HNOB	Heme NO-binding domain
HNOX	Heme NO- and oxygen-binding domain
HTH	Helix-turn-helix domain
HOLI	Ligand-binding domain of hormone receptors
Hpk2	Histidine protein kinase 2 from *Treponema denticola*
YC-1	5-[1-(phenylmethyl)- 1H-indazol-3-yl]-2-furanmethanol
LBD	Ligand-binding domain of the nuclear receptor
LOV	Light–oxygen–voltage-sensing domain
MRI	Magnetic resonance imaging
Mtb	*Mycobacterium tuberculosis*
NifA	Oxygen-responsive nitrogen fixation sensor protein A
NorR	NO reductase regulatory protein
NosP	Nitric oxide-sensing protein
PAS	Per, period circadian protein, Arnt, aryl hydrocarbon receptor nuclear translocator protein, Sim, single-minded protein domain
pGpG	Linearized form of cyclic di-GMP, 5′-phosphoguanylyl- (3′ −> 5′)-guanosine
SAXS	Small-angle X-ray scattering
SCHIC	Sensor-containing heme instead of cobalamin domain
sGC	Soluble guanylate cyclase
ThkA	*Thermotoga maritima* histidine kinase A
*Tt*H-NOX	H-NOX sensor from *Thermoanaerobacter tengcongensis*
VcBhr-DGC	*Vibrio cholerae* bacterial hemerythrin diguanylate cyclase sensor protein
Whib	regulator originally associated to the gene locus involved with the conversion of white multinucleoidal aseptate aerial hyphae into chains of mature grey uninucleoidal spores
Whib3	NO and/or O_2 sensor from *Mycobacterium tuberculosis*
YFP	Yellow fluorescent protein

References

1. Gonzaga de França Lopes, L.; Gouveia Júnior, F.S.; Karine Medeiros Holanda, A.; Maria Moreira de Carvalho, I.; Longhinotti, E.; Paulo, T.F.; Abreu, D.S.; Bernhardt, P.V.; Gilles-Gonzalez, M.-A.; Cirino Nogueira Diógenes, I.; et al. Bioinorganic systems responsive to the diatomic gases O_2, NO, and CO: From biological sensors to therapy. *Coord. Chem. Rev.* **2021**, *445*, 214096. [CrossRef]
2. Negrerie, M. Iron transitions during activation of allosteric heme proteins in cell signaling. *Metallomics* **2019**, *11*, 868–893. [CrossRef]
3. Gilles-Gonzalez, M.A.; Gonzalez, G. Heme-based sensors: Defining characteristics, recent developments, and regulatory hypotheses. *J. Inorg. Biochem.* **2005**, *99*, 1–22. [CrossRef] [PubMed]
4. Shimizu, T.; Huang, D.; Yan, F.; Stranava, M.; Bartosova, M.; Fojtikova, V.; Martinkova, M. Gaseous O_2, NO, and CO in signal transduction: Structure and function relationships of heme-based gas sensors and heme-redox sensors. *Chem. Rev.* **2015**, *115*, 6491–6533. [CrossRef] [PubMed]
5. Gilles-Gonzalez, M.-A.; Gonzalez, G. A Surfeit of Biological Heme-based Sensors. In *The Smallest Biomolecules: Diatomics and Their Interactions with Heme Proteins*; Ghosh, A., Ed.; Elsevier: Amsterdam, The Netherlands, 2008.
6. Britt, R.D.; Rao, G.; Tao, L. Bioassembly of complex iron-sulfur enzymes: Hydrogenases and nitrogenases. *Nat. Rev. Chem.* **2020**, *4*, 542–549. [CrossRef] [PubMed]
7. D'Autreaux, B.; Tucker, N.P.; Dixon, R.; Spiro, S. A non-haem iron centre in the transcription factor NorR senses nitric oxide. *Nature* **2005**, *437*, 769–772. [CrossRef]
8. Schaller, R.A.; Ali, S.K.; Klose, K.E.; Kurtz, D.M., Jr. A bacterial hemerythrin domain regulates the activity of a *Vibrio cholerae* diguanylate cyclase. *Biochemistry* **2012**, *51*, 8563–8570. [CrossRef]
9. Crack, J.C.; Green, J.; Thomson, A.J.; Le Brun, N.E. Iron-sulfur clusters as biological sensors: The chemistry of reactions with molecular oxygen and nitric oxide. *Acc. Chem. Res.* **2014**, *47*, 3196–3205. [CrossRef]
10. Gilles-Gonzalez, M.A.; Ditta, G.S.; Helinski, D.R. A haemoprotein with kinase activity encoded by the oxygen sensor of *Rhizobium meliloti*. *Nature* **1991**, *350*, 170–172. [CrossRef]

11. Hossain, S.; Boon, E.M. Discovery of a Novel Nitric Oxide Binding Protein and Nitric-Oxide-Responsive Signaling Pathway in *Pseudomonas aeruginosa*. *ACS Infect. Dis.* **2017**, *3*, 454–461. [CrossRef]

12. Taabazuing, C.Y.; Hangasky, J.A.; Knapp, M.J. Oxygen sensing strategies in mammals and bacteria. *J. Inorg. Biochem.* **2014**, *133*, 63–72. [CrossRef]

13. Fandrey, J.; Schodel, J.; Eckardt, K.U.; Katschinski, D.M.; Wenger, R.H. Now a Nobel gas: Oxygen. *Pflug. Arch.* **2019**, *471*, 1343–1358. [CrossRef] [PubMed]

14. Crack, J.C.; Thomson, A.J.; Le Brun, N.E. Mass spectrometric identification of intermediates in the O_2-driven [4Fe-4S] to [2Fe-2S] cluster conversion in FNR. *Proc. Natl. Acad. Sci. USA* **2017**, *114*, E3215–E3223. [CrossRef] [PubMed]

15. Gilles-Gonzalez, M.A.; Gonzalez, G.; Sousa, E.H.; Tuckerman, J. Oxygen-sensing histidine-protein kinases: Assays of ligand binding and turnover of response-regulator substrates. *Methods Enzymol.* **2008**, *437*, 173–189. [CrossRef]

16. Fojtikova, V.; Stranava, M.; Vos, M.H.; Liebl, U.; Hranicek, J.; Kitanishi, K.; Shimizu, T.; Martinkova, M. Kinetic Analysis of a Globin-Coupled Histidine Kinase, AfGcHK: Effects of the Heme Iron Complex, Response Regulator, and Metal Cations on Autophosphorylation Activity. *Biochemistry* **2015**, *54*, 5017–5029. [CrossRef] [PubMed]

17. Arnold, W.P.; Mittal, C.K.; Katsuki, S.; Murad, F. Nitric oxide activates guanylate cyclase and increases guanosine $3':5'$-cyclic monophosphate levels in various tissue preparations. *Proc. Natl. Acad. Sci. USA* **1977**, *74*, 3203–3207. [CrossRef]

18. Patterson, D.C.; Ruiz, M.P.; Yoon, H.; Walker, J.A.; Armache, J.P.; Yennawar, N.H.; Weinert, E.E. Differential ligand-selective control of opposing enzymatic activities within a bifunctional c-di-GMP enzyme. *Proc. Natl. Acad. Sci. USA* **2021**, *118*, e2100657118. [CrossRef] [PubMed]

19. Tuckerman, J.R.; Gonzalez, G.; Sousa, E.H.; Wan, X.; Saito, J.A.; Alam, M.; Gilles-Gonzalez, M.A. An oxygen-sensing diguanylate cyclase and phosphodiesterase couple for c-di-GMP control. *Biochemistry* **2009**, *48*, 9764–9774. [CrossRef]

20. Shimizu, T.; Lengalova, A.; Martinek, V.; Martinkova, M. Heme: Emergent roles of heme in signal transduction, functional regulation and as catalytic centres. *Chem. Soc. Rev.* **2019**, *48*, 5624–5657. [CrossRef]

21. Gilles-Gonzalez, M.A.; Sousa, E.H.S. Escherichia coli DosC and DosP: A role of c-di-GMP in compartmentalized sensing by degradosomes. *Adv. Microb. Physiol.* **2019**, *75*, 53–67. [CrossRef]

22. Tuckerman, J.R.; Gonzalez, G.; Gilles-Gonzalez, M.A. Cyclic di-GMP activation of polynucleotide phosphorylase signal-dependent RNA processing. *J. Mol. Biol.* **2011**, *407*, 633–639. [CrossRef]

23. Shelver, D.; Kerby, R.L.; He, Y.; Roberts, G.P. CooA, a CO-sensing transcription factor from *Rhodospirillum rubrum*, is a CO-binding heme protein. *Proc. Natl. Acad. Sci. USA* **1997**, *94*, 11216–11220. [CrossRef]

24. Schofield, C.J.; Ratcliffe, P.J. Oxygen sensing by HIF hydroxylases. *Nat. Rev. Mol. Cell Biol.* **2004**, *5*, 343–354. [CrossRef]

25. Nunes, E.D.; Villela, A.D.; Basso, L.A.; Teixeira, E.H.; Andrade, A.L.; Vasconcelos, M.A.; Neto, L.G.D.; Gondim, A.C.S.; Diogenes, I.C.N.; Romo, A.I.B.; et al. Light-induced disruption of an acyl hydrazone link as a novel strategy for drug release and activation: Isoniazid as a proof-of-concept case. *Inorg. Chem. Front.* **2020**, *7*, 859–870. [CrossRef]

26. Kang, Y.; Liu, R.; Wu, J.X.; Chen, L. Structural insights into the mechanism of human soluble guanylate cyclase. *Nature* **2019**, *574*, 206–210. [CrossRef]

27. Wittenborn, E.C.; Marletta, M.A. Structural Perspectives on the Mechanism of Soluble Guanylate Cyclase Activation. *Int. J. Mol. Sci.* **2021**, *22*, 5439. [CrossRef]

28. Horst, B.G.; Yokom, A.L.; Rosenberg, D.J.; Morris, K.L.; Hammel, M.; Hurley, J.H.; Marletta, M.A. Allosteric activation of the nitric oxide receptor soluble guanylate cyclase mapped by cryo-electron microscopy. *Elife* **2019**, *8*, e50634. [CrossRef]

29. Lanzilotta, W.N.; Schuller, D.J.; Thorsteinsson, M.V.; Kerby, R.L.; Roberts, G.P.; Poulos, T.L. Structure of the CO sensing transcription activator CooA. *Nat. Struct. Biol.* **2000**, *7*, 876–880. [CrossRef]

30. Williams, D.E.; Nisbett, L.M.; Bacon, B.; Boon, E. Bacterial Heme-Based Sensors of Nitric Oxide. *Antioxid. Redox Signal.* **2018**, *29*, 1872–1887. [CrossRef]

31. Stuehr, D.J.; Misra, S.; Dai, Y.; Ghosh, A. Maturation, inactivation, and recovery mechanisms of soluble guanylyl cyclase. *J. Biol. Chem.* **2021**, *296*, 100336. [CrossRef]

32. Friebe, A.; Sandner, P.; Schmidtko, A. cGMP: A unique 2nd messenger molecule—Recent developments in cGMP research and development. *Naunyn-Schmiedebergs Arch. Pharmacol.* **2020**, *393*, 287–302. [CrossRef]

33. Sousa, E.H.S.; Gonzalez, G.; Gilles-Gonzalez, M.A. Soluble guanylate cyclase and its microbial relatives. In *The Smallest Biomolecules: Perspectives on Heme-Diatomic Interactions*; Ghosh, A., Ed.; Elsevier: Amsterdam, The Netherlands, 2008; pp. 524–539.

34. Bian, K.; Murad, F. What is next in nitric oxide research? From cardiovascular system to cancer biology. *Nitric Oxide* **2014**, *43*, 3–7. [CrossRef]

35. Rutten, P.J.; Poole, P.S. Oxygen regulatory mechanisms of nitrogen fixation in rhizobia. *Adv. Microb. Physiol.* **2019**, *75*, 325–389. [CrossRef]

36. Sousa, E.H.S.; Gilles-Gonzalez, M.A. Haem-Based Sensors of O2: Lessons and Perspectives. *Adv. Microb. Physiol.* **2017**, *71*, 235–257. [CrossRef]

37. Ganesh, I.; Gwon, D.A.; Lee, J.W. Gas-Sensing Transcriptional Regulators. *Biotechnol. J.* **2020**, *15*, e1900345. [CrossRef]

38. Dixon, R.; Kahn, D. Genetic regulation of biological nitrogen fixation. *Nat. Rev. Microbiol.* **2004**, *2*, 621–631. [CrossRef]

39. Ledermann, R.; Schulte, C.C.M.; Poole, P.S. How Rhizobia Adapt to the Nodule Environment. *J. Bacteriol.* **2021**, *203*, e0053920. [CrossRef]

40. Wright, G.S.A.; Saeki, A.; Hikima, T.; Nishizono, Y.; Hisano, T.; Kamaya, M.; Nukina, K.; Nishitani, H.; Nakamura, H.; Yamamoto, M.; et al. Architecture of the complete oxygen-sensing FixL-FixJ two-component signal transduction system. *Sci. Signal.* **2018**, *11*. [CrossRef]

41. Gilles-Gonzalez, M.A.; Caceres, A.I.; Sousa, E.H.; Tomchick, D.R.; Brautigam, C.; Gonzalez, C.; Machius, M. A proximal arginine R206 participates in switching of the *Bradyrhizobium japonicum* FixL oxygen sensor. *J. Mol. Biol.* **2006**, *360*, 80–89. [CrossRef]

42. Balland, V.; Bouzhir-Sima, L.; Anxolabéhère-Mallart, E.; Boussac, A.; Vos, M.H.; Liebl, U.; Mattioli, T.A. Functional Implications of the Propionate 7−Arginine 220 Interaction in the FixLH Oxygen Sensor from *Bradyrhizobium japonicum*. *Biochemistry* **2006**, *45*, 2072–2084. [CrossRef]

43. Sousa, E.H.; Gonzalez, G.; Gilles-Gonzalez, M.A. Oxygen blocks the reaction of the FixL-FixJ complex with ATP but does not influence binding of FixJ or ATP to FixL. *Biochemistry* **2005**, *44*, 15359–15365. [CrossRef]

44. Key, J.; Moffat, K. Crystal structures of deoxy and CO-bound bjFixLH reveal details of ligand recognition and signaling. *Biochemistry* **2005**, *44*, 4627–4635. [CrossRef]

45. Tuckerman, J.R.; Gonzalez, G.; Dioum, E.M.; Gilles-Gonzalez, M.A. Ligand and oxidation-state specific regulation of the heme-based oxygen sensor FixL from *Sinorhizobium meliloti*. *Biochemistry* **2002**, *41*, 6170–6177. [CrossRef]

46. Hao, B.; Isaza, C.; Arndt, J.; Soltis, M.; Chan, M.K. Structure-based mechanism of O2 sensing and ligand discrimination by the FixL heme domain of *Bradyrhizobium japonicum*. *Biochemistry* **2002**, *41*, 12952–12958. [CrossRef]

47. Gong, W.; Hao, B.; Mansy, S.S.; Gonzalez, G.; Gilles-Gonzalez, M.A.; Chan, M.K. Structure of a biological oxygen sensor: A new mechanism for heme-driven signal transduction. *Proc. Natl. Acad. Sci. USA* **1998**, *95*, 15177–15182. [CrossRef]

48. Guimaraes, W.G.; Gondim, A.C.S.; Costa, P.; Gilles-Gonzalez, M.A.; Lopes, L.G.F.; Carepo, M.S.P.; Sousa, E.H.S. Insights into signal transduction by a hybrid FixL: Denaturation study of on and off states of a multi-domain oxygen sensor. *J. Inorg. Biochem.* **2017**, *172*, 129–137. [CrossRef]

49. Honorio-Felicio, N.; Carepo, M.S.; Paulo, T.D.F.; de Franca Lopes, L.G.; Sousa, E.H.; Diogenes, I.C.; Bernhardt, P.V. The Heme-Based Oxygen Sensor *Rhizobium etli* FixL: Influence of Auxiliary Ligands on Heme Redox Potential and Implications on the Enzyme Activity. *J. Inorg. Biochem.* **2016**, *164*, 34–41. [CrossRef]

50. Yamawaki, T.; Ishikawa, H.; Mizuno, M.; Nakamura, H.; Shiro, Y.; Mizutani, Y. Regulatory Implications of Structural Changes in Tyr201 of the Oxygen Sensor Protein FixL. *Biochemistry* **2016**, *55*, 4027–4035. [CrossRef]

51. Yamada, S.; Sugimoto, H.; Kobayashi, M.; Ohno, A.; Nakamura, H.; Shiro, Y. Structure of PAS-linked histidine kinase and the response regulator complex. *Structure* **2009**, *17*, 1333–1344. [CrossRef]

52. Miksovska, J.; Suquet, C.; Satterlee, J.D.; Larsen, R.W. Characterization of conformational changes coupled to ligand photodissoci-ation from the heme binding domain of FixL. *Biochemistry* **2005**, *44*, 10028–10036. [CrossRef]

53. Dunham, C.M.; Dioum, E.M.; Tuckerman, J.R.; Gonzalez, G.; Scott, W.G.; Gilles-Gonzalez, M.A. A distal arginine in oxygen-sensing heme-PAS domains is essential to ligand binding, signal transduction, and structure. *Biochemistry* **2003**, *42*, 7701–7708. [CrossRef]

54. Miyatake, H.; Mukai, M.; Adachi, S.; Nakamura, H.; Tamura, K.; Iizuka, T.; Shiro, Y.; Strange, R.W.; Hasnain, S.S. Iron coordination structures of oxygen sensor FixL characterized by Fe K-edge extended X-ray absorption fine structure and resonance raman spectroscopy. *J. Biol. Chem.* **1999**, *274*, 23176–23184. [CrossRef]

55. Gong, W.; Hao, B.; Chan, M.K. New mechanistic insights from structural studies of the oxygen-sensing domain of *Bradyrhizobium japonicum* FixL. *Biochemistry* **2000**, *39*, 3955–3962. [CrossRef] [PubMed]

56. Capra, E.J.; Laub, M.T. Evolution of two-component signal transduction systems. *Annu. Rev. Microbiol.* **2012**, *66*, 325–347. [CrossRef] [PubMed]

57. Sousa, E.H.; Tuckerman, J.R.; Gondim, A.C.; Gonzalez, G.; Gilles-Gonzalez, M.A. Signal transduction and phosphoryl transfer by a FixL hybrid kinase with low oxygen affinity: Importance of the vicinal PAS domain and receiver aspartate. *Biochemistry* **2013**, *52*, 456–465. [CrossRef]

58. Gilles-Gonzalez, M.A.; Gonzalez, G.; Perutz, M.F.; Kiger, L.; Marden, M.C.; Poyart, C. Heme-based sensors, exemplified by the kinase FixL, are a new class of heme protein with distinctive ligand binding and autoxidation. *Biochemistry* **1994**, *33*, 8067–8073. [CrossRef]

59. Sousa, E.H.; Tuckerman, J.R.; Gonzalez, G.; Gilles-Gonzalez, M.A. A memory of oxygen binding explains the dose response of the heme-based sensor FixL. *Biochemistry* **2007**, *46*, 6249–6257. [CrossRef]

60. Vauquelin, G.; Charlton, S.J. Long-lasting target binding and rebinding as mechanisms to prolong in vivo drug action. *Br. J. Pharmacol.* **2010**, *161*, 488–508. [CrossRef]

61. Nuernberger, P.; Lee, K.F.; Bonvalet, A.; Bouzhir-Sima, L.; Lambry, J.C.; Liebl, U.; Joffre, M.; Vos, M.H. Strong ligand-protein interactions revealed by ultrafast infrared spectroscopy of CO in the heme pocket of the oxygen sensor FixL. *J. Am. Chem. Soc.* **2011**, *133*, 17110–17113. [CrossRef] [PubMed]

62. Key, J.; Srajer, V.; Pahl, R.; Moffat, K. Time-resolved crystallographic studies of the heme domain of the oxygen sensor FixL: Structural dynamics of ligand rebinding and their relation to signal transduction. *Biochemistry* **2007**, *46*, 4706–4715. [CrossRef]

63. Yano, S.; Ishikawa, H.; Mizuno, M.; Nakamura, H.; Shiro, Y.; Mizutani, Y. Ultraviolet resonance Raman observations of the structural dynamics of rhizobial oxygen sensor FixL on ligand recognition. *J. Phys. Chem. B* **2013**, *117*, 15786–15791. [CrossRef]

64. Yamawaki, T.; Mizuno, M.; Ishikawa, H.; Takemura, K.; Kitao, A.; Shiro, Y.; Mizutani, Y. Regulatory Switching by Concerted Motions on the Microsecond Time Scale of the Oxygen Sensor Protein FixL. *J. Phys. Chem. B* **2021**. [CrossRef]

65. Crosson, S.; McGrath, P.T.; Stephens, C.; McAdams, H.H.; Shapiro, L. Conserved modular design of an oxygen sensory/signaling network with species-specific output. *Proc. Natl. Acad. Sci. USA* **2005**, *102*, 8018–8023. [CrossRef] [PubMed]

66. Roset, M.S.; Almiron, M.A. FixL-like sensor FlbS of Brucella abortus binds haem and is necessary for survival within eukaryotic cells. *FEBS Lett.* **2013**, *587*, 3102–3107. [CrossRef] [PubMed]

67. Murthy, U.M.; Wecker, M.S.; Posewitz, M.C.; Gilles-Gonzalez, M.A.; Ghirardi, M.L. Novel FixL homologues in *Chlamydomonas reinhardtii* bind heme and O(2). *FEBS Lett.* **2012**, *586*, 4282–4288. [CrossRef]

68. Schaefers, M.M.; Liao, T.L.; Boisvert, N.M.; Roux, D.; Yoder-Himes, D.; Priebe, G.P. An Oxygen-Sensing Two-Component System in the Burkholderia cepacia Complex Regulates Biofilm, Intracellular Invasion, and Pathogenicity. *PLoS Pathog.* **2017**, *13*, e1006116. [CrossRef] [PubMed]

69. Schaefers, M.M.; Wang, B.X.; Boisvert, N.M.; Martini, S.J.; Bonney, S.L.; Marshall, C.W.; Laub, M.T.; Cooper, V.S.; Priebe, G.P. Evolution towards Virulence in a Burkholderia Two-Component System. *MBio* **2021**, *12*, e0182321. [CrossRef] [PubMed]

70. WHO. *Global Tuberculosis Report*; World Health Organization: Geneva, Switzerland, 2020.

71. Campanico, A.; Harjivan, S.G.; Warner, D.F.; Moreira, R.; Lopes, F. Addressing Latent Tuberculosis: New Advances in Mimicking the Disease, Discovering Key Targets, and Designing Hit Compounds. *Int. J. Mol. Sci.* **2020**, *21*, 8854. [CrossRef] [PubMed]

72. Sousa, E.H.S.; Diogenes, I.C.N.; Lopes, L.G.F.; Moura, J.J.G. Potential therapeutic approaches for a sleeping pathogen: Tuberculosis a case for bioinorganic chemistry. *J. Biol. Inorg. Chem.* **2020**, *25*, 685–704. [CrossRef]

73. Dasgupta, N.; Kapur, V.; Singh, K.K.; Das, T.K.; Sachdeva, S.; Jyothisri, K.; Tyagi, J.S. Characterization of a two-component system, devR-devS, of *Mycobacterium tuberculosis*. *Tuber. Lung Dis.* **2000**, *80*, 141–159. [CrossRef]

74. Saini, D.K.; Malhotra, V.; Dey, D.; Pant, N.; Das, T.K.; Tyagi, J.S. DevR-DevS is a bona fide two-component system of *Mycobacterium tuberculosis* that is hypoxia-responsive in the absence of the DNA-binding domain of DevR. *Microbiology* **2004**, *150*, 865–875. [CrossRef] [PubMed]

75. Saini, D.K.; Malhotra, V.; Tyagi, J.S. Cross talk between DevS sensor kinase homologue, Rv2027c, and DevR response regulator of *Mycobacterium tuberculosis*. *FEBS Lett.* **2004**, *565*, 75–80. [CrossRef] [PubMed]

76. Roberts, D.M.; Liao, R.P.; Wisedchaisri, G.; Hol, W.G.; Sherman, D.R. Two sensor kinases contribute to the hypoxic response of *Mycobacterium tuberculosis*. *J. Biol. Chem.* **2004**, *279*, 23082–23087. [CrossRef]

77. Gupta, V.K.; Kumar, M.M.; Singh, D.; Bisht, D.; Sharma, S. Drug targets in dormant *Mycobacterium tuberculosis*: Can the conquest against tuberculosis become a reality? *Infect. Dis.* **2018**, *50*, 81–94. [CrossRef]

78. Sardiwal, S.; Kendall, S.L.; Movahedzadeh, F.; Rison, S.C.; Stoker, N.G.; Djordjevic, S. A GAF domain in the hypoxia/NO-inducible *Mycobacterium tuberculosis* DosS protein binds haem. *J. Mol. Biol.* **2005**, *353*, 929–936. [CrossRef] [PubMed]

79. Sousa, E.H.S.; Tuckerman, J.R.; Gonzalez, G.; Gilles-Gonzalez, M.-A. DosT and DevS are oxygen-switched kinases in *Mycobacterium tuberculosis*. *Protein Sci.* **2007**, *16*, 1708–1719. [CrossRef]

80. Sousa, E.H.S.; Gonzalez, G.; Gilles-Gonzalez, M.A. Target DNA stabilizes *Mycobacterium tuberculosis* DevR/DosR phosphorylation by the full-length oxygen sensors DevS/DosS and DosT. *FEBS J.* **2017**, *284*, 3954–3967. [CrossRef]

81. Kumar, A.; Toledo, J.C.; Patel, R.P.; Lancaster, J.R.; Steyn, A.J.C. *Mycobacterium tuberculosis* DosS is a redox sensor and DosT is a hypoxia sensor. *Proc. Natl. Acad. Sci. USA* **2007**, *104*, 11568–11573. [CrossRef] [PubMed]

82. Barreto, G.A.; Carepo, M.S.P.; Gondim, A.C.S.; Guimarães, W.G.; Lopes, L.G.F.; Bernhardt, P.V.; Paulo, T.F.; Sousa, E.H.S.; Diógenes, I.C.N. A spectroelectrochemical investigation of the heme-based sensor DevS from *Mycobacterium tuberculosis*: A redox versus oxygen sensor. *FEBS J.* **2019**, *286*, 4278–4293. [CrossRef]

83. Honaker, R.W.; Leistikow, R.L.; Bartek, I.L.; Voskuil, M.I. Unique roles of DosT and DosS in DosR regulon induction and *Mycobacterium tuberculosis* dormancy. *Infect. Immun.* **2009**, *77*, 3258–3263. [CrossRef]

84. Cho, H.Y.; Cho, H.J.; Kim, Y.M.; Oh, J.I.; Kang, B.S. Structural insight into the heme-based redox sensing by DosS from *Mycobacterium tuberculosis*. *J. Biol. Chem.* **2009**, *284*, 13057–13067. [CrossRef]

85. Ioanoviciu, A.; Meharenna, Y.T.; Poulos, T.L.; Ortiz de Montellano, P.R. DevS oxy complex stability identifies this heme protein as a gas sensor in *Mycobacterium tuberculosis* dormancy. *Biochemistry* **2009**, *48*, 5839–5848. [CrossRef]

86. Podust, L.M.; Ioanoviciu, A.; Ortiz de Montellano, P.R. 2.3 A X-ray structure of the heme-bound GAF domain of sensory histidine kinase DosT of *Mycobacterium tuberculosis*. *Biochemistry* **2008**, *47*, 12523–12531. [CrossRef] [PubMed]

87. Basudhar, D.; Madrona, Y.; Yukl, E.T.; Sivaramakrishnan, S.; Nishida, C.R.; Moenne-Loccoz, P.; Ortiz de Montellano, P.R. Distal Hydrogen-bonding Interactions in Ligand Sensing and Signaling by *Mycobacterium tuberculosis* DosS. *J. Biol. Chem.* **2016**, *291*, 16100–16111. [CrossRef]

88. Yukl, E.T.; Ioanoviciu, A.; de Montellano, P.R.; Moenne-Loccoz, P. Interdomain interactions within the two-component heme-based sensor DevS from *Mycobacterium tuberculosis*. *Biochemistry* **2007**, *46*, 9728–9736. [CrossRef] [PubMed]

89. Lobao, J.; Gondim, A.C.S.; Guimaraes, W.G.; Gilles-Gonzalez, M.A.; Lopes, L.G.F.; Sousa, E.H.S. Oxygen triggers signal transduction in the DevS (DosS) sensor of *Mycobacterium tuberculosis* by modulating the quaternary structure. *FEBS J.* **2019**, *286*, 479–494. [CrossRef]

90. Skalova, T.; Lengalova, A.; Dohnalek, J.; Harlos, K.; Mihalcin, P.; Kolenko, P.; Stranava, M.; Blaha, J.; Shimizu, T.; Martinkova, M. Disruption of the dimerization interface of the sensing domain in the dimeric heme-based oxygen sensor AfGcHK abolishes bacterial signal transduction. *J. Biol. Chem.* **2020**, *295*, 1587–1597. [CrossRef]

91. Burns, J.L.; Rivera, S.; Deer, D.D.; Joynt, S.C.; Dvorak, D.; Weinert, E.E. Oxygen and Bis(3′,5′)-cyclic Dimeric Guanosine Monophosphate Binding Control Oligomerization State Equilibria of Diguanylate Cyclase-Containing Globin Coupled Sensors. *Biochemistry* **2016**, *55*, 6642–6651. [CrossRef] [PubMed]

92. Sousa, E.H.; Lopes, L.G.; Gonzalez, G.; Gilles-Gonzalez, M.A. Drug discovery targeting heme-based sensors and their coupled activities. *J. Inorg. Biochem.* **2017**, *167*, 12–20. [CrossRef]

93. Sandner, P.; Zimmer, D.P.; Milne, G.T.; Follmann, M.; Hobbs, A.; Stasch, J.P. Soluble Guanylate Cyclase Stimulators and Activators. *Handb. Exp. Pharmacol.* **2021**, *264*, 355–394. [CrossRef] [PubMed]

94. Muniz Carvalho, E.; Silva Sousa, E.H.; Bernardes-Génisson, V.; Gonzaga de França Lopes, L. When NO Is not Enough: Chemical Systems, Advances and Challenges in the Development of NO and HNO Donors for Old and Current Medical Issues. *Eur. J. Inorg. Chem.* **2021**, *2021*, 4316–4348. [CrossRef]

95. Ko, F.N.; Wu, C.C.; Kuo, S.C.; Lee, F.Y.; Teng, C.M. YC-1, a novel activator of platelet guanylate cyclase. *Blood* **1994**, *84*, 4226–4233. [CrossRef]

96. Sa, D.S.; Fernandes, A.F.; Silva, C.D.; Costa, P.P.; Fonteles, M.C.; Nascimento, N.R.; Lopes, L.G.; Sousa, E.H. Non-nitric oxide based metallovasodilators: Synthesis, reactivity and biological studies. *Dalton Trans.* **2015**, *44*, 13633–13640. [CrossRef]

97. He, B.; Chen, Z. Molecular Targets for Small-Molecule Modulators of Circadian Clocks. *Curr. Drug Metab.* **2016**, *17*, 503–512. [CrossRef] [PubMed]

98. Ellermann, M.; Sperandio, V. Bacterial signaling as an antimicrobial target. *Curr. Opin. Microbiol.* **2020**, *57*, 78–86. [CrossRef] [PubMed]

99. Wille, J.; Coenye, T. Biofilm dispersion: The key to biofilm eradication or opening Pandora's box? *Biofilm* **2020**, *2*, 100027. [CrossRef]

100. Cai, Y.M.; Webb, J.S. Optimization of nitric oxide donors for investigating biofilm dispersal response in *Pseudomonas aeruginosa* clinical isolates. *Appl. Microbiol. Biotechnol.* **2020**, *104*, 8859–8869. [CrossRef] [PubMed]

101. da Silva Filho, P.M.; Andrade, A.L.; Lopes, J.; Pinheiro, A.A.; de Vasconcelos, M.A.; Fonseca, S.; Lopes, L.G.F.; Sousa, E.H.S.; Teixeira, E.H.; Longhinotti, E. The biofilm inhibition activity of a NO donor nanosilica with enhanced antibiotics action. *Int. J. Pharm.* **2021**, *610*, 121220. [CrossRef]

102. Howlin, R.P.; Cathie, K.; Hall-Stoodley, L.; Cornelius, V.; Duignan, C.; Allan, R.N.; Fernandez, B.O.; Barraud, N.; Bruce, K.D.; Jefferies, J.; et al. Low-Dose Nitric Oxide as Targeted Anti-biofilm Adjunctive Therapy to Treat Chronic *Pseudomonas aeruginosa* Infection in Cystic Fibrosis. *Mol. Ther.* **2017**, *25*, 2104–2116. [CrossRef] [PubMed]

103. Boce, M.; Tasse, M.; Mallet-Ladeira, S.; Pillet, F.; Da Silva, C.; Vicendo, P.; Lacroix, P.G.; Malfant, I.; Rols, M.P. Effect of trans(NO, OH)-[RuFT(Cl)(OH)NO](PF6) ruthenium nitrosyl complex on methicillin-resistant *Staphylococcus epidermidis*. *Sci. Rep.* **2019**, *9*, 4867. [CrossRef] [PubMed]

104. Zheng, H.; Abramovitch, R.B. Inhibiting DosRST as a new approach to tuberculosis therapy. *Future Med. Chem.* **2020**, *12*, 457–467. [CrossRef]

105. Gupta, R.K.; Thakur, T.S.; Desiraju, G.R.; Tyagi, J.S. Structure-based design of DevR inhibitor active against nonreplicating *Mycobacterium tuberculosis*. *J. Med. Chem.* **2009**, *52*, 6324–6334. [CrossRef]

106. Kaur, K.; Taneja, N.K.; Dhingra, S.; Tyagi, J.S. DevR (DosR) mimetic peptides impair transcriptional regulation and survival of *Mycobacterium tuberculosis* under hypoxia by inhibiting the autokinase activity of DevS sensor kinase. *BMC Microbiol.* **2014**, *14*, 195. [CrossRef] [PubMed]

107. Dhingra, S.; Kaur, K.; Taneja, N.K.; Tyagi, J.S. DevR (DosR) binding peptide inhibits adaptation of *Mycobacterium tuberculosis* under hypoxia. *FEMS Microbiol. Lett.* **2012**, *330*, 66–71. [CrossRef]

108. Zheng, H.; Colvin, C.J.; Johnson, B.K.; Kirchhoff, P.D.; Wilson, M.; Jorgensen-Muga, K.; Larsen, S.D.; Abramovitch, R.B. Inhibitors of *Mycobacterium tuberculosis* DosRST signaling and persistence. *Nat. Chem. Biol.* **2017**, *13*, 218–225. [CrossRef]

109. Zheng, H.; Williams, J.T.; Aleiwi, B.; Ellsworth, E.; Abramovitch, R.B. Inhibiting *Mycobacterium tuberculosis* DosRST Signaling by Targeting Response Regulator DNA Binding and Sensor Kinase Heme. *ACS Chem. Biol.* **2020**, *15*, 52–62. [CrossRef]

110. Patel, Y.S.; Mistry, N.; Mehra, S. Repurposing artemisinin as an anti-mycobacterial agent in synergy with rifampicin. *Tuberculosis* **2019**, *115*, 146–153. [CrossRef] [PubMed]

111. Choi, W.H. Novel Pharmacological Activity of Artesunate and Artemisinin: Their Potential as Anti-Tubercular Agents. *J. Clin. Med.* **2017**, *6*, 30. [CrossRef]

112. Wan, X.; Tuckerman, J.R.; Saito, J.A.; Freitas, T.A.; Newhouse, J.S.; Denery, J.R.; Galperin, M.Y.; Gonzalez, G.; Gilles-Gonzalez, M.A.; Alam, M. Globins synthesize the second messenger bis-(3′-5′)-cyclic diguanosine monophosphate in bacteria. *J. Mol. Biol.* **2009**, *388*, 262–270. [CrossRef]

113. Sarkar, J.; Miller, D.P.; Oliver, L.D.; Marconi, R.T. The Treponema denticola PAS Domain-Containing Histidine Kinase Hpk2 Is a Heme Binding Sensor of Oxygen Levels. *J. Bacteriol.* **2018**, *200*, e00116–e00118. [CrossRef]

114. Sen Santara, S.; Roy, J.; Mukherjee, S.; Bose, M.; Saha, R.; Adak, S. Globin-coupled heme containing oxygen sensor soluble adenylate cyclase in Leishmania prevents cell death during hypoxia. *Proc. Natl. Acad. Sci. USA* **2013**, *110*, 16790–16795. [CrossRef]

115. Abuchowski, A. SANGUINATE (PEGylated Carboxyhemoglobin Bovine): Mechanism of Action and Clinical Update. *Artif. Organs* **2017**, *41*, 346–350. [CrossRef]

116. Azarov, I.; Wang, L.; Rose, J.J.; Xu, Q.; Huang, X.N.; Belanger, A.; Wang, Y.; Guo, L.; Liu, C.; Ucer, K.B.; et al. Five-coordinate H64Q neuroglobin as a ligand-trap antidote for carbon monoxide poisoning. *Sci. Transl. Med.* **2016**, *8*, 368ra173. [CrossRef]

117. Hoops, H.E.; Manning, J.E.; Graham, T.L.; McCully, B.H.; McCurdy, S.L.; Ross, J.D. Selective aortic arch perfusion with fresh whole blood or HBOC-201 reverses hemorrhage-induced traumatic cardiac arrest in a lethal model of noncompressible torso hemorrhage. *J. Trauma Acute Care Surg.* **2019**, *87*, 263–273. [CrossRef]

118. Alam, M.; Larsen, R. Heme Proteins hemAT-Hs and hemAT-Bs and Their Use in Medicine and Microsensors. U.S. Patent 7,129,329, 31 October 2006.

119. Boehme, J.; Le Moan, N.; Kameny, R.J.; Loucks, A.; Johengen, M.J.; Lesneski, A.L.; Gong, W.; Goudy, B.D.; Davis, T.; Tanaka, K.; et al. Preservation of myocardial contractility during acute hypoxia with OMX-CV, a novel oxygen delivery biotherapeutic. *PLoS Biol.* **2018**, *16*, e2005924. [CrossRef]

120. Itoh, T.; Matsuura, S.I.; Chuong, T.T.; Tanaike, O.; Hamakawa, S.; Shimizu, T. Successful Mesoporous Silica Encapsulation of Optimally Functional EcDOS (*E. coli* Direct Oxygen Sensor), a Heme-based O_2-Sensing Phosphodiesterase. *Anal. Sci.* **2019**, *35*, 329–335. [CrossRef]

121. Kalia, D.; Merey, G.; Nakayama, S.; Zheng, Y.; Zhou, J.; Luo, Y.; Guo, M.; Roembke, B.T.; Sintim, H.O. Nucleotide, c-di-GMP, c-di-AMP, cGMP, cAMP, (p)ppGpp signaling in bacteria and implications in pathogenesis. *Chem. Soc. Rev.* **2013**, *42*, 305–341. [CrossRef]

122. Opoku-Temeng, C.; Sintim, H.O. Targeting c-di-GMP Signaling, Biofilm Formation, and Bacterial Motility with Small Molecules. *Methods Mol. Biol.* **2017**, *1657*, 419–430. [CrossRef]

123. Wang, J.; Karpus, J.; Zhao, B.S.; Luo, Z.; Chen, P.R.; He, C. A selective fluorescent probe for carbon monoxide imaging in living cells. *Angew. Chem. Int. Ed. Engl.* **2012**, *51*, 9652–9656. [CrossRef]

124. Winter, M.B.; McLaurin, E.J.; Reece, S.Y.; Olea, C., Jr.; Nocera, D.G.; Marletta, M.A. Ru-porphyrin protein scaffolds for sensing O_2. *J. Am. Chem. Soc.* **2010**, *132*, 5582–5583. [CrossRef]

125. Nierth, A.; Marletta, M.A. Direct meso-Alkynylation of Metalloporphyrins Through Gold Catalysis for Hemoprotein Engineering. *Angew. Chem. Int. Ed.* **2014**, *53*, 2611–2614. [CrossRef]

126. Lemon, C.M.; Marletta, M.A. Corrole-Substituted Fluorescent Heme Proteins. *Inorg. Chem.* **2021**, *60*, 2716–2729. [CrossRef] [PubMed]

127. Fardeau, M.L.; Salinas, M.B.; L'Haridon, S.; Jeanthon, C.; Verhe, F.; Cayol, J.L.; Patel, B.K.C.; Garcia, J.L.; Ollivier, B. Isolation from oil reservoirs of novel thermophilic anaerobes phylogenetically related to *Thermoanaerobacter subterraneus*: Reassignment of *T. subterraneus*, *Thermoanaerobacter yonseiensis*, *Thermoanaerobacter tengcongensis* and *Carboxydibrachium pacificum* to *Caldanaerobacter subterraneus* gen. nov., sp. nov., comb. nov. as four novel subspecies. *Int. J. Syst. Evol. Microbiol.* **2004**, *54*, 467–474. [CrossRef] [PubMed]

128. Dai, Z.; Boon, E.M. Engineering of the heme pocket of an H-NOX domain for direct cyanide detection and quantification. *J. Am. Chem. Soc.* **2010**, *132*, 11496–11503. [CrossRef]

129. Winter, M.B.; Klemm, P.J.; Phillips-Piro, C.M.; Raymond, K.N.; Marletta, M.A. Porphyrin-substituted H-NOX proteins as high-relaxivity MRI contrast agents. *Inorg. Chem.* **2013**, *52*, 2277–2279. [CrossRef]

130. Kumita, H.; Yamada, S.; Nakamura, H.; Shiro, Y. Chimeric sensory kinases containing O_2 sensor domain of FixL and histidine kinase domain from thermophile. *Biochim. Biophys. Acta* **2003**, *1646*, 136–144. [CrossRef]

131. Moglich, A.; Ayers, R.A.; Moffat, K. Design and signaling mechanism of light-regulated histidine kinases. *J. Mol. Biol.* **2009**, *385*, 1433–1444. [CrossRef]

132. Iniesta, A.A.; Hillson, N.J.; Shapiro, L. Cell pole-specific activation of a critical bacterial cell cycle kinase. *Proc. Natl. Acad. Sci. USA* **2010**, *107*, 7012–7017. [CrossRef]

133. Nomata, J.; Hisabori, T. Development of heme protein based oxygen sensing indicators. *Sci. Rep.* **2018**, *8*, 11849. [CrossRef]

134. Ainala, S.K.; Seol, E.; Sekar, B.S.; Park, S. Improvement of carbon monoxide-dependent hydrogen production activity in Citrobacter amalonaticus Y19 by over-expressing the CO-sensing transcriptional activator, CooA. *Int. J. Hydrog. Energ.* **2014**, *39*, 10417–10425. [CrossRef]

135. Arnold, F.H. Innovation by Evolution: Bringing New Chemistry to Life (Nobel Lecture). *Angew. Chem. Int. Ed. Engl.* **2019**, *58*, 14420–14426. [CrossRef]

136. Leveson-Gower, R.B.; Mayer, C.; Roelfes, G. The importance of catalytic promiscuity for enzyme design and evolution. *Nat. Rev. Chem.* **2019**, *3*, 687–705. [CrossRef]

137. Aggrey-Fynn, J.E.; Surmeli, N.B. A novel thermophilic hemoprotein scaffold for rational design of biocatalysts. *J. Biol. Inorg. Chem.* **2018**, *23*, 1295–1307. [CrossRef]

138. Ortmayer, M.; Lafite, P.; Menon, B.R.; Tralau, T.; Fisher, K.; Denkhaus, L.; Scrutton, N.S.; Rigby, S.E.; Munro, A.W.; Hay, S.; et al. An oxidative N-demethylase reveals PAS transition from ubiquitous sensor to enzyme. *Nature* **2016**, *539*, 593–597. [CrossRef] [PubMed]

139. Woodward, J.J.; Martin, N.I.; Marletta, M.A. An Escherichia coli expression-based method for heme substitution. *Nat. Methods* **2007**, *4*, 43–45. [CrossRef] [PubMed]

140. Kawakami, N.; Shoji, O.; Watanabe, Y. Single-step reconstitution of apo-hemoproteins at the disruption stage of Escherichia coli cells. *Chembiochem* **2012**, *13*, 2045–2047. [CrossRef] [PubMed]

141. Lelyveld, V.S.; Brustad, E.; Arnold, F.H.; Jasanoff, A. Metal-substituted protein MRI contrast agents engineered for enhanced relaxivity and ligand sensitivity. *J. Am. Chem. Soc.* **2011**, *133*, 649–651. [CrossRef]
142. Perkins, L.J.; Weaver, B.R.; Buller, A.R.; Burstyn, J.N. De novo biosynthesis of a nonnatural cobalt porphyrin cofactor in *E. coli* and incorporation into hemoproteins. *Proc. Natl. Acad. Sci. USA* **2021**, *118*, e2017625118. [CrossRef]

Review

Translating Biochemistry Concepts into Cartoons and Graphic Narratives: Potential and Pitfalls

Mireia Alemany-Pagès [1], Rui Tavares [1,2], Anabela Marisa Azul [3] and João Ramalho-Santos [4,*]

1   CNC—Center for Neuroscience and Cell Biology, CIBB, Rua Larga, University of Coimbra, 3004-504 Coimbra, Portugal; mireia.apages@gmail.com (M.A.-P.); ruidiastavares@gmail.com (R.T.)
2   PhD Programme in Experimental Biology and Biomedicine (PDBEB), IIIUC—Institute for Interdisciplinary Research, University of Coimbra, 3030-789 Coimbra, Portugal
3   IIIUC—Institute for Interdisciplinary Research, University of Coimbra, 3030-789 Coimbra, Portugal; amjrazul@ci.uc.pt
4   DCV—Department of Life Sciences, University of Coimbra, Calçada Martim de Freitas, 3000-456 Coimbra, Portugal
*   Correspondence: jramalho@uc.pt

Abstract: Simple biochemical concepts can be hard to grasp by non-specialists, even when they are related to practical contexts in industry, day-to-day activities, or well-acknowledged pathological conditions. This is especially important in instances where accurate communication of biochemical aspects for different types of stakeholders may be crucial. Examples include interacting with policymakers to establish guidelines, with patients (and/or caregivers) to identify key concepts in promoting awareness and adherence to therapeutic regimens, or with teachers and students for novel approaches in critical thinking. Focusing on our own work in developing communication tools for different purposes, in this review we will focus on some examples of how biochemical concepts can be effectively translated into illustrations and graphical narratives. For this purpose, engagement with target audiences in developing the materials themselves is key. We also discuss how specific projects can be tailored for different purposes, as well as evidence that comic-book strategies are effective in conveying biochemical and biomedical knowledge.

Keywords: metabolism; comics; biomedical knowledge; health communication; science communication

Citation: Alemany-Pagès, M.; Tavares, R.; Azul, A.M.; Ramalho-Santos, J. Translating Biochemistry Concepts into Cartoons and Graphic Narratives: Potential and Pitfalls. *BioChem* **2022**, *2*, 104–114. https://doi.org/10.3390/biochem2010008

Academic Editor: Buyong Ma

Received: 24 November 2021
Accepted: 9 March 2022
Published: 17 March 2022

Publisher's Note: MDPI stays neutral with regard to jurisdictional claims in published maps and institutional affiliations.

1. Introduction

Scientific concepts in Biochemistry, and the life sciences in general, can be hard to understand, and the use of images to better convey information is common at all levels, from very basic leaflets tailored to a lay audience to schoolbooks and the most advanced scientific papers [1–6]. In fact, no Biochemistry or Molecular Biology textbook is devoid of images (photos, graphs, diagrams, and schematics) to help guide the reader [7,8].

In many cases, sequential images are used to display a sequence of events, for example relating to enzyme-mediated catalysis, the binding of ligands to receptors, signaling cascades, or steps in embryo development. One of the challenges in both education and outreach towards different types of audiences is translating biochemical concepts to a target audience that may need assistance in grasping the most basic information. While classical textbook images are effective at more professional levels, they may not be the ideal choice to communicate with all audiences.

It is therefore unsurprising to note that science communication aiming at different target audiences often uses illustrations, sketch notes, cartoons, and comics, with or without an accompanying narrative [9–15]. In this review, we will discuss examples of how biochemical concepts can be effectively translated into graphical narratives, providing tools for both formal and informal education, as well as for more effective stakeholder engagement. Importantly, we will also discuss the appropriate methodology needed to create effective comics-based materials. The considerations to take into account include content selection, graphic considerations regarding style and representation, preparatory interactions with a putative target audience to tailor the content to specific needs, and the monitoring of impact to ensure that key messages are indeed transmitted. To the best of our knowledge, this is the first review to simultaneously focus on all these aspects of comics-based science communication.

2. Using Sequential Images and Comics in Science Communication

By combining words and images, comics, in particular, embody a reliable format to effectively communicate biomedical knowledge [16] as well as for introducing scientific information in the classroom [12,17]. Moreover, using narrative techniques and well-designed characters and narratives can also potentiate effectiveness in terms of scientific or health-related beliefs, attitudes, intentions, and behaviors [18–22]. Although some understanding of the specificities and conventions of the language is needed, comics seem to be approachable and able to reach and engage new audiences, also showing that entertainment and education are not mutually exclusive [23]. On the other hand, the narrative dimension contextualizes the scientific or health-related knowledge, both through verbal and visual cues, and engages the reader at a personal level [14,16,23]. Importantly comics have been clearly shown to be a successful tool in science communication, in some cases outperforming other forms of communication dealing with the same content (text, animations, etc.) [17,23–27].

In a biomedical context, examples include efforts dedicated to human immunod-eficiency virus (HIV)/AIDS prevention programs [28,29], but later also sexual health education [30–32], rheumatoid disease [33,34], cancer [35–37], mental health [38–43] and metabolic disorders [44–47]. More recently, and concomitant with a gradual transition from hard copies to digital media, several comics designed to promote healthier diets, increased physical activity, and weight loss as preventive and therapeutic practices for obesity, diabetes, and cardiovascular diseases have been produced [48–53].

3. Defining Representations, Topics, Characters, and Storylines

Several aspects can be considered when using comics or visual-related language via a series of illustrations. These must be carefully considered and tailored towards the specific goals of a given project, starting with the choices to show scientific content in a simplified (not simplistic) manner.

For example, the use of analogies or metaphors is pervasive throughout scientific discourse in general and to convey biochemical concepts in particular. As shown in Figure 1, ATP production in the mitochondria follows a well-understood process that involves electron transport through several complexes coupled with proton pumping across the inner mitochondrial membrane. The resulting electrochemical gradient (mitochondrial membrane potential) is then used by the ATP synthase to phosphorylate ADP to ATP. At the molecular level, the mechanism regarding ATP synthase activity is often represented as rotating paddles converting different forms of energy. On the other hand, to convey the basic notions of what is at stake, the mitochondrial membrane potential can be metaphorically represented as a hydroelectric dam and the organelle itself as a battery (Figure 1C). Although these metaphors or analogies can also lead to misconceptions and improper simplifications, they can nevertheless be useful, especially as a first approach or in contacting audiences with limited scientific knowledge. In another example, discussing the molecular basis of actin-myosin interactions leading to voluntary skeletal muscle contraction in sarcomeres can be done logically in the context of exercise (possibly also related to the promotion of healthy lifestyles), for example, using the similarities with rowing as a metaphor (Figure 2).

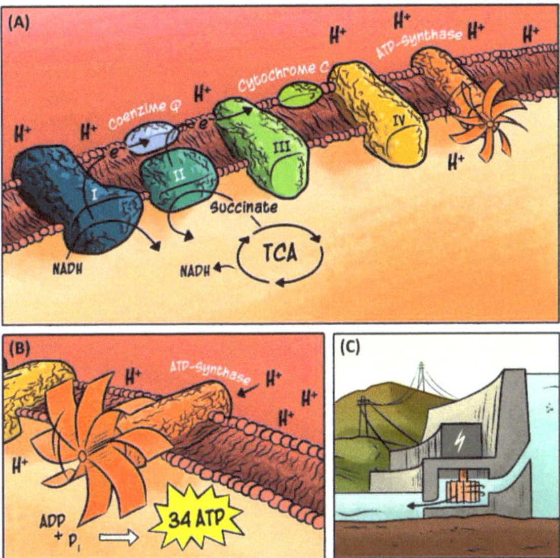

Figure 1. Typical representation of the electron transport chain in the inner mitochondrial membrane and the generation of an electrochemical gradient (**A**). The use of the gradient to synthesize ATP by the ATP synthase can be depicted as a system of rotating paddles based on structural information (**B**). More metaphorical is an analogy to show how a hydroelectric dam produced electricity (**C**).

Figure 2. The introduction of a common activity (rowing) to stress the movements of myosin and actin filaments relevant in muscular contraction.

Another use of images to convey biochemical concepts includes the recently established Graphic Medicine genre, which applies to comics paradigms and principles generated with Narrative Medicine. This approach aims to humanize medical care, focusing on the experiences of those involved (patients, practitioners, family members, caregivers, health care systems, community) [54–59], stemming from individual experiences [60–63]. However, while the stories may resonate with others under the same circumstances, their goal is not necessarily to convey biochemical aspects related to a specific pathology or explain a treatment regimen from a scientific/clinical standpoint. Therefore, in a more educational context, this type of strategy may not be useful, as it introduces subjectivity and can neglect paramount scientific information [55,57].

A similar discussion may be had in terms of visual style [64,65], as both realistic depictions and cartoonish representations are possible. Indeed they are often combined, for example, using human characters as patients and cartoons to anthropomorphize phenomena (such as cancer, for example), cells, or proteins, thus equipping them with a clearer sense of (metaphorical) agency [14,16,66–68]. The goal is to engage readers using different strategies, leading to very different tools. The issues at stake are evident in three comics that focus on DNA and its functions. While the subject matter is essentially the same and accurately depicted in each case, one [69] opts for a more realistic style characteristic of classical textbooks (drawn photos of key researchers and detailed schematics of molecular events), another relies on the anthropomorphizing of important molecules (helicases, transcription factors, repair enzymes) to better represent their actions [70]. Yet another has human characters interact directly with molecular aspects [71]. These choices may be related to the different goals of each approach: a thorough historical narrative on the discovery of DNA, an accessible approach on DNA function for the general population, and a textbook in Japanese comic (mangá) format to engage young readers, respectively. Regardless of the choices made, the images should not distract the reader from the main message. This can be the case with more expressionist styles drawings featuring unclear or unnecessary details, a confusing color palette, over-complex visual analogies, or even page layouts with an unintuitive reading experience.

On the other hand, the choices of what to communicate and how should actively involve the potential end-users, in other words, adopt a participatory approach in co-producing the content [72]. One has to start by identifying the main topics to communicate and the target audience, and then try to work with said audience (in focus groups or short interviews, preferably with some open questions) to establish a realistic knowledge baseline [73–76]. A few questions may need to be answered, for example: what is the background knowledge? In material focused on a specific medical issue, what do patients know about their own condition? What do learners know about the topic? What are the gaps in knowledge? What are audiences curious about knowing? Could it make a difference in how they interact with a given technology, adherence to treatment, or lifestyle changes? What do practitioners think would be useful information for patients (and caregivers) to have in order to elicit behavioral changes? What do teachers and professors think would be useful for the students' perception and cognition?

Patients and other learners, even those with limited literacy, have their own experiences and may have more developed and detailed knowledge than should be assumed, although misconceptions are also common [14,16]. These mutual interactions may also be useful if the goal is to create characters and narratives. If, for example, characters in an educational comic, as well as their biographies/behaviors, are based on a composite of different members of a target audience, there might be a better chance of other readers from the same population identifying with the fictional characters. One example is shown in Figure 3 and discussed in detail below [77].

It is important to note that this should be as open and as collaborative a process as possible, although financial, organizational, and time constraints will likely play a role. Ideally, an initial effort should lead to a first draft that can then be shared for possible feedback and eventually result in the refining of both visual and verbal narratives [75,76,78–80]. Did the audience understand the key concepts and message accurately? Did they appreciate the comic or think the story, characters, and script were convincing? Did they find something off-putting, and why? Finally, distribution of the final product should not be considered as the ending, but rather an evaluation of impact should be performed, if possible. Again, the goal, in this case, is to determine how effective the comic (or whatever other tool) was in a wider context, or if, for instance, its effectiveness varied with different segments of the population (age, gender, culture, literacy levels). These two aspects can be carried out using appropriately structured questionnaires (or interviews), which must nevertheless be validated using appropriate methodologies and expertise [26,77].

Figure 3. The use of characters with specific biographies and storylines to help convey both biochemical information (the synthesis of fat from sugars) and to help convey information relative to nutrition and healthier lifestyles (see text for discussion).

4. Comics and the Communicating Biochemical Concepts in a Context of Metabolic Disorders

Biochemical aspects related to metabolic disorders are a good place to start, as they can be important in many relevant biomedical and educational contexts [38,81]. Previously we faced the tensions of discussing accuracy versus aesthetics in our work on metabolic disorders focusing on mitochondrial biology [47]. Is there a realistic limit for depicting cellular mechanisms in a graphic narrative? What were the best strategies to bring mitochondria to 'real-life', convey the fundamental research, and engage in healthier habits? The narrative is about a young woman diagnosed with a metabolic disorder, and transports the reader to a realistic view of mitochondrial biology blended with the changing daily routines of the character. The intent was to target different types of lay audiences, not only patients but also children or at-risk segments, and students at different levels of education. This is relevant both in conditions that are relatively well-known, such as diabetes, as well as conditions where that is not at all the case, such as Nonalcoholic Fatty Liver Disease (NAFLD). NAFLD has been the target of a few campaigns exactly because it is not only underdiagnosed but a burden predicted to increase, especially in the developed world [77].

But, besides the obvious healthy lifestyle aspects related to both prevention and management of metabolic disorders (proper nutrition, regular physical activity, and exercise) as well as regular monitoring (blood sugar and cholesterol levels, liver enzyme activity etc.), what biochemical aspects can be conveyed to strengthen the message? In our work focusing on type 2 diabetes patients at high risk for NAFLD [82], a few issues arose out of conversations with researchers, practitioners, and patients who were interviewed for the project, albeit at different levels [83]. Two of these issues can be cited because they involve clear mechanistic biochemical insights: Why do people who avoid fatty foods (but not sweets) have fat accumulation and obesity-related problems (de novo lipogenesis)? Why does insulin sometimes "stop working" (insulin resistance)?

The advantage of excess sugars being converted to fat is clear from the standpoint of efficient energy storage, but conveying even a simple version of the biochemical reactions involved would likely confuse more than enlighten a lay audience [84]. One possible solution (Figure 3) is using a well-known nutritional choice (the ingestion of a sugary beverage) in parallel with a simple schematic showing that the liver can convert sugar to fat. The narrative, in this case, involves an older relative explaining to a child why drinking that particular beverage is, in a biochemical sense, equivalent to eating fatty foods [85,86]. The older character functions both as an explainer and a positive role model [20], and it should be noted that this particular character is aware of these issues because she suffers from type 2 diabetes. She is, in fact, a composite character based on the biographies and experiences of some of the patients interviewed at the start of the project [83]. This is an example of how more abstract notions may be interspersed with real-life situations different readers can relate to.

On the other hand, the concept of insulin resistance is crucial to explain when interacting with a lay audience [87,88]. First of all, the normal situation regarding the function of pancreas-produced insulin to ensure proper blood sugar management must be explained before its deregulation can be addressed. In this case, a possible solution is using cartoons that embody cells and molecules with anthropomorphic characteristics, as well as with an omnipotent narrator solution [82]. After explaining how the process usually works, insulin resistance is then represented as a systems overload that, by burdening biochemical circuits with an excess activity it can no longer cope with, causes a "rebellion" or a "strike", leading to several important health-related consequences (Figure 4).

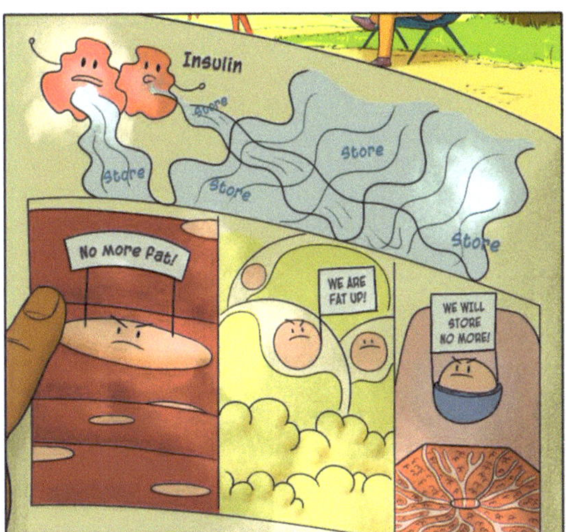

Figure 4. Anthropomorphizing biochemical elements such as molecules (Insulin) or cells/tissues (from left to right: muscle fat, liver) as a communication tool, in this case to convey the concept of insulin resistance using a "cells on strike" metaphor. See text for discussion.

5. Planning a Project: Monitoring Impact and Tailoring Strategies in Graphic Narrative-Based Science Communication

Although comics-based science communication is not new, what is novel is the increasing interest in scholarly research in this field, involving different types of analyses [89–92]. In addition, an important issue is whether concepts are in fact learned by a target audience. Importantly, comics have been shown to be effective in this regard in different contexts, notably in the classroom [12,42]. In our work, we previously showed that comics were particularly effective compared to other methods in terms of conveying meaning and complex biomedical concepts to diverse audiences. Specifically, readers with little previous knowledge of the basic characteristics and biomedical applications of stem cells could grasp complex notions (pluripotency, multipotency, differentiation, reprogramming) by interacting with a tailored comic [26]. The methodology to assess this knowledge increase used structured questionnaires given to participants before and after interacting with different science communication materials on the same topics (texts, interviews, and animations, besides the comic). However, this initial effort had two clear limitations. On the one hand, the materials had been produced by basic researchers without any interaction with target audiences to pre-validate the different approaches or introduce any clarifying changes. On the other hand, impact monitoring was planned after the materials were produced, not as an integral part of the project. It is extremely important that besides the issues already discussed (choosing a type of narrative and graphic style), both issues should be considered when planning a project.

In our current work on metabolic disorders, and as previously stated, we interacted with patients [83] to better tailor the comic book produced [82]. Additionally, we also planned to assess impact using two structured questionnaires before and after interaction with the comic. While the data obtained suggest that the comic was indeed effective in both increasing awareness related to NAFLD, explaining the biochemistry behind the pathology, and transmitting key concepts such as insulin resistance (Alemany-Pagès et al., submitted), it is worth noting that the COVID-19 pandemic greatly limited contact with users of an online platform, thus selecting a specific segment of the possible target audience.

Additionally, although the biochemical aspects may be the same regardless of context, the context does matter when communicating effectively. Besides an understandable and approachable scientific content, if a message is to be received and acted upon, the audience must be able to fully identify with the situations depicted. These should not seem "foreign" to them [14,16,20,74]. For example, when fictional elements are depicted related to everyday life situations involving human characters, the types of people represented, landscape, housing, clothing, social dynamics, eating habits, environment, etc., must be considered [74,89]. In this case, one size might not fit all, and the same materials might have to be changed to reach different audiences in different contexts (social backgrounds, countries, local environment) effectively [77].

In conclusion, although more structured and planned projects meeting the criteria noted above are certainly needed, comics-based narratives seem to be a valid tool in the field of science communication, specifically in transmitting biochemical and biomedical concepts.

Author Contributions: Conceptualization, J.R.-S.; critical literature search and analysis, M.A.-P.; figures and graphic considerations, R.T.; writing—original draft preparation J.R.-S.; writing—editing and review M.A.-P., A.M.A., J.R.-S. All authors have read and agreed to the published version of the manuscript.

Funding: This study received support from the FOIE GRAS project, funded by the European Union's Horizon 2020, Research and Innovation programme under the Marie Skłodowska-Curie Grant Agreement No. 722619 and the European Regional Development Fund (ERDF), through the COMPETE 2020—Operational Programme for Competitiveness and Internationalisation and Portuguese national funds via FCT—Fundação para a Ciência e a Tecnologia, under the projects UIDB/04539/2020, UIDP/04539/2020 and LA/P/0058/2020; and through the Centro 2020 Regional Operational Programme: project CENTRO-01-0145-FEDER-000012-HealthyAging2020.

Institutional Review Board Statement: Not applicable.

Informed Consent Statement: Not applicable.

Conflicts of Interest: The authors declare no conflict of interest.

References

1. Theel, E.S.; McAdam, A.J. What about Serology? A Micro-Comic Strip. *J. Clin. Microbiol.* **2019**, *57*, e00797-19. [CrossRef] [PubMed]
2. Hosler, J. Science Comics. Available online: http://jayhosler.com/science-comics.html (accessed on 19 November 2021).
3. Hosler, J. *Optical Allusions*; CreateSpace Independent Publishing Platform: Scotts Valley, CA, USA, 2013; ISBN 978-1482387773.
4. Delp, C.; Jones, J. Communicating information to patients: The use of cartoon illustrations to improve comprehension of instructions. *Acad. Emerg. Med.* **1996**, *3*, 264–270. [CrossRef] [PubMed]
5. Houts, P.S.; Doak, C.C.; Doak, L.G.; Loscalzo, M.J. The role of pictures in improving health communication: A review of research on attention, comprehension, recall, and adherence. *Patient Educ. Couns.* **2006**, *61*, 173–190. [CrossRef]
6. Zehr, E.P. From Claude Bernard to the Batcave and beyond: Using Batman as a hook for physiology education. *Adv. Physiol. Educ.* **2011**, *35*, 1–4. [CrossRef] [PubMed]
7. Nelson, D.L.; Cox, M.M. *Lehninger Principles of Biochemistry*, 8th ed.; W.H. Freeman: New York, NY, USA, 2021.
8. Alberts, B.; Heald, R.; Johnson, A.; Lewis, J.; Morgan, D.; Raff, M.C.; Roberts, K.; Walter, P. *Molecular Biology of the Cell*, 7th ed.; W.W. Norton & Company: New York, NY, USA, 2022.
9. Barry, A.M.; Gazzaniga, M. Science and Visual Communication. 2008; pp. 1–11. Available online: https://www.giantscreencinema.com/Portals/0/BarryPaperFinal.pdf (accessed on 19 November 2021).
10. Flemming, D.; Cress, U.; Kimmig, S.; Brandt, M.; Kimmerle, J. Emotionalization in Science Communication: The Impact of Narratives and Visual Representations on Knowledge Gain and Risk Perception. *Front. Commun.* **2018**, *3*, 3. [CrossRef]
11. Bucchi, M. *Science and the Media: Alternative Routes in Scientific Communication*; Routledge: London, UK, 1998.
12. Tatalovic, M. Science comics as tools for science education and communication: A brief, exploratory study. *SISSA Int. Sch. Adv. Stud. J. Sci. Commun.* **2009**, *8*, 1824–2049. [CrossRef]
13. Tribull, C.M. Sequential Science: A Guide to Communication Through Comics. *Ann. Entomol. Soc. Am.* **2017**, *110*, 457–466. [CrossRef]
14. Farinella, M. The potential of comics in science communication. *J. Sci. Commun.* **2018**, *17*, Y01. [CrossRef]
15. Fernández-Fontecha, A.; O'Halloran, K.L.; Tan, S.; Wignell, P. A multimodal approach to visual thinking: The scientific sketchnote. *Vis. Commun.* **2018**, *18*, 5–29. [CrossRef]
16. Jee, B.D.; Anggoro, F.K. Comic cognition: Exploring the potential cognitive impacts of science comics. *J. Cogn. Educ. Psychol.* **2012**, *11*, 196–208. [CrossRef]
17. Aleixo, P.A.; Sumner, K. Memory for biopsychology material presented in comic book format. *J. Graph. Nov. Comics* **2017**, *8*, 79–88. [CrossRef]

18. Shen, F.; Sheer, V.C.; Li, R. Impact of Narratives on Persuasion in Health Communication: A Meta-Analysis. *J. Advert.* **2015**, *44*, 105–113. [CrossRef]

19. Petraglia, J. Narrative intervention in behavior and public health. *J. Health Commun.* **2007**, *12*, 493–505. [CrossRef]

20. Slater, M.D. Entertainment education and the persuasive impact of narratives. In *Narrative Impact: Social and Cognitive Foundations*; Green, M.C., Strange, J.J., Brock, T.C., Eds.; Lawrence Erlbaum Associates Publishers: Mahwah, NJ, USA, 2002; pp. 157–181.

21. Hinyard, L.J.; Kreuter, M.W. Using narrative communication as a tool for health behavior change: A conceptual, theoretical, and empirical overview. *Health Educ. Behav.* **2007**, *34*, 777–792. [CrossRef]

22. Greenhalgh, T. *Cultural Contexts of Health: The Use of Narrative Research in the Health Sector*; World Health Organization: Geneva, Switzerland, 2016.

23. Collver, J.; Weitkamp, E. Alter egos: An exploration of the perspectives and identities of science comic creators. *J. Sci. Commun.* **2018**, *17*, A01. [CrossRef]

24. Hosler, J.; Boomer, K.B. Are comic books an effective way to engage nonmajors in learning and appreciating science? *CBE Life Sci. Educ.* **2011**, *10*, 309–317. [CrossRef]

25. Spiegel, A.N.; Mcquillan, J.; Halpin, P.; Matuk, C.; Diamond, J. Engaging Teenagers with Science Through Comics. *Res. Sci. Educ.* **2013**, *43*, 2309–2326. [CrossRef]

26. Amaral, S.V.; Forte, T.; Ramalho-Santos, J.; Da Cruz, M.T.G. I want more and better cells!—An outreach project about stem cells and its impact on the general population. *PLoS ONE* **2015**, *10*, e0133753. [CrossRef]

27. Pratt, H.J. Narrative in Comics. *J. Aesthet. Art Crit.* **2009**, *67*, 107–117. [CrossRef]

28. Cain, B. Saying it with feeling: Photonovels and comic books in development. *Dev. Commun. Rep.* **1986**, *55*, 1–2.

29. Pillai, V.K.; Kelley, A.C. Men and family planning: Toward a policy of male involvement. *Pol. Popul. Rev.* **1994**, *5*, 293–304.

30. Garbarino, J. Children's response to a sexual abuse prevention program: A study of the Spiderman comic. *Child Abus. Negl.* **1987**, *11*, 143–148. [CrossRef]

31. Speizer, I.S.; Calhoun, L.M.; Guilkey, D.K. Reaching Urban Female Adolescents at Key Points of Sexual and Reproductive Health Transitions: Evidence from a Longitudinal Study from Kenya. *Afr. J. Reprod. Health* **2018**, *22*, 47–59. [PubMed]

32. Jacoby, S.D.; Lucarelli, M.; Musse, F.; Krishnamurthy, A.; Salyers, V. A Mixed-Methods Study of Immigrant Somali Women's Health Literacy and Perinatal Experiences in Maine. *J. Midwifery Women's Health* **2015**, *60*, 593–603. [CrossRef]

33. Massone, F.; Martínez, M.E.; Pascual-Ramos, V.; Quintana, R.; Stange, L.; Caballero-Uribe, C.V.; Massardo, L. Educational website incorporating rheumatoid arthritis patient needs for Latin American and Caribbean countries. *Clin. Rheumatol.* **2017**, *36*, 2789–2797. [CrossRef] [PubMed]

34. Mendelson, A.; Rabinowicz, N.; Reis, Y.; Amarilyo, G.; Harel, L.; Hashkes, P.J.; Uziel, Y. Comics as an educational tool for children with juvenile idiopathic arthritis. *Pediatric Rheumatol.* **2017**, *15*, 69. [CrossRef] [PubMed]

35. Putnam, G.L.; Yanagisako, K.L. Skin cancer comic book: Evaluation of a public educational vehicle. *J. Audiov. Media Med.* **1985**, *8*, 22–25. [CrossRef]

36. Alam, S.; Elwyn, G.; Percac-Lima, S.; Grande, S.; Durand, M.-A. Assessing the acceptability and feasibility of encounter decision aids for early stage breast cancer targeted at underserved patients. *BMC Med. Inform. Decis. Mak.* **2016**, *16*, 147. [CrossRef] [PubMed]

37. Criado, P.R.; Ocampo-Garza, J.; Brasil, A.L.D.; Belda Junior, W.; Di Chiacchio, N.; de Moraes, A.M.; Vasconcellos, C. Skin cancer prevention campaign in childhood: Survey based on 3676 children in Brazil. *J. Eur. Acad. Dermatol. Venereol.* **2018**, *32*, 1272–1277. [CrossRef]

38. King, A.J. Using Comics to Communicate About Health: An Introduction to the Symposium on Visual Narratives and Graphic Medicine. *Health Commun.* **2017**, *32*, 523–524. [CrossRef]

39. Shirotsuki, K.; Nonaka, Y.; Takano, J.; Abe, K.; Adachi, S.-I.; Adachi, S.; Nakao, M. Brief internet-based cognitive behavior therapy program with a supplement drink improved anxiety and somatic symptoms in Japanese workers. *Biopsychosoc. Med.* **2017**, *11*, 25. [CrossRef]

40. Imamura, K.; Kawakami, N.; Furukawa, T.A.; Matsuyama, Y.; Shimazu, A.; Umanodan, R.; Kasai, K. Effects of an internet-based cognitive behavioral therapy intervention on improving work engagement and other work-related outcomes: An analysis of secondary outcomes of a randomized controlled trial. *J. Occup. Environ. Med.* **2015**, *57*, 578–584. [CrossRef] [PubMed]

41. Tekle-Haimanot, R.; Preux, P.M.; Gerard, D.; Worku, D.K.; Belay, H.D.; Gebrewold, M.A. Impact of an educational comic book on epilepsy-related knowledge, awareness, and attitudes among school children in Ethiopia. *Epilepsy Behav.* **2016**, *61*, 218–223. [CrossRef]

42. Cicero, C.E.; Giuliano, L.; Todaro, V.; Colli, C.; Padilla, S.; Vilte, E.; Nicoletti, A. Comic book-based educational program on epilepsy for high-school students: Results from a pilot study in the Gran Chaco region, Bolivia. *Epilepsy Behav.* **2020**, *107*, 107076. [CrossRef] [PubMed]

43. Hernandez, M.Y.; Organista, K.C. Entertainment-education? A fotonovela? A new strategy to improve depression literacy and help-seeking behaviors in at-risk immigrant Latinas. *Am. J. Community Psychol.* **2013**, *52*, 224–235. [CrossRef] [PubMed]

44. Leung, M.M.; Tripicchio, G.; Agaronov, A. Manga Comic Influences Snack Selection in Black and Hispanic New York City Youth. *J. Nutr. Educ. Behav.* **2014**, *46*, 142–147. [CrossRef] [PubMed]

45. Tarver, T.; Woodson, D.; Fechter, N.; Vanchiere, J.; Olmstadt, W.; Tudor, C. A Novel Tool for Health Literacy: Using Comic Books to Combat Childhood Obesity. *J. Hosp. Librariansh.* **2016**, *16*, 152–159. [CrossRef] [PubMed]

46. Thompson, D.; Mahabir, R.; Bhatt, R.; Boutte, C.; Cantu, D.; Vazquez, I.; Buday, R. Butterfly Girls; promoting healthy diet and physical activity to young African American girls online: Rationale and design. *BMC Public Health* **2013**, *13*, 709. [CrossRef]

47. Azul, A.M.; Ramalho-Santos, J.; Oliveira, P.J.; Tavares, R. Mitochondrial Follies: A Short Journey in Life and Energy. In *Mitochondrial Biology and Experimental Therapeutics*; Oliveira, P.J., Ed.; Springer International Publishing: Cham, Germany, 2018; pp. 649–692. [CrossRef]

48. Ko, L.K.; Rillamas-Sun, E.; Bishop, S.; Cisneros, O.; Holte, S.; Thompson, B. Together We STRIDE: A quasi-experimental trial testing the effectiveness of a multi-level obesity intervention for Hispanic children in rural communities. *Contemp. Clin. Trials* **2018**, *67*, 81–86. [CrossRef]

49. Matsuzono, K.; Yokota, C.; Takekawa, H.; Okamura, T.; Miyamatsu, N.; Nakayama, H.; Watanabe, T. Effects of stroke education of junior high school students on stroke knowledge of their parents: Tochigi project. *Stroke* **2015**, *46*, 572–574. [CrossRef] [PubMed]

50. Kato, S.; Okamura, T.; Kuwabara, K.; Takekawa, H.; Nagao, M.; Umesawa, M.; Minematsu, K. Effects of a school-based stroke education program on stroke-related knowledge and behaviour modification-school class based intervention study for elementary school students and parental guardians in a Japanese rural area. *BMJ Open* **2017**, *7*, e017632. [CrossRef]

51. Ohyama, S.; Yokota, C.; Miyashita, F.; Amano, T.; Inoue, Y.; Shigehatake, Y.; Minematsu, K. Effective Education Materials to Advance Stroke Awareness Without Teacher Participation in Junior High School Students. *J. Stroke Cerebrovasc. Dis.* **2015**, *24*, 2533–2538. [CrossRef] [PubMed]

52. Leung, A.Y.M.; Chau, P.H.; Leung, I.S.H.; Tse, M.; Wong, P.L.C.; Tam, W.M.; Leung, D.Y. Motivating Diabetic and Hypertensive Patients to Engage in Regular Physical Activity: A Multi-Component Intervention Derived from the Concept of Photovoice. *Int. J. Environ. Res. Public Health* **2019**, *16*, 1219. [CrossRef] [PubMed]

53. Learning about Diabetes. 2020. Available online: https://learningaboutdiabetes.org/programs-consumer/ (accessed on 19 November 2021).

54. Charon, R. Medicine, the novel, and the passage of time. *Ann. Intern. Med.* **2000**, *132*, 63–68. [CrossRef] [PubMed]

55. Charon, R. Narrative Medicine: Form, Function, and Ethics. *Ann. Intern. Med.* **2001**, *134*, 83–87. [CrossRef]

56. Green, M.J.; Czerwiec, M.K. Graphic Medicine: The Best of 2016. *JAMA* **2016**, *316*, 2580–2581. [CrossRef]

57. Williams, I.; Squier, S.; Myers, K.S. *Graphic Medicine Manifesto*; Penn State University Press: University Park, PA, USA, 2015; ISBN 978-0-271-06649-3.

58. Myers, K.R.; George, D.R.; Huang, X.; Goldenberg, M.D.F.; Van Scoy, L.J.; Lehman, E.; Green, M.J. Use of a Graphic Memoir to Enhance Clinicians' Understanding of and Empathy for Patients with Parkinson Disease. *Perm. J.* **2020**, *24*, 19.060. [CrossRef]

59. Czerwiec, M.K. Representing AIDS in Comics. *AMA J. Ethics* **2018**, *20*, 199–205.

60. Czerwiec, M.K. *Taking Turns: Stories from HIV/AIDS Care Unit 371*; Penn State University Press: University Park, PA, USA, 2017; ISBN 978-0271078182.

61. Forney, E. *Marbles: Mania, Depression, Michelangelo & Me: A Graphic Memoir*; Robinson: London, UK, 2013; ISBN 978-1592407323.

62. Fies, B. *Mom's Cancer*; Abrams ComicArts: New York, NY, USA, 2011; ISBN 978-0810971073.

63. Marchetto, M.A. *Cancer Vixen*; Pantheon: New York, NY, USA, 2009; ISBN 978-0375714740.

64. Waite, M. Writing medical comics. *J. Vis. Commun. Med.* **2019**, *42*, 144–150. [CrossRef] [PubMed]

65. Walker, S. Effective antimicrobial resistance communication: The role of information design. *Palgrave Commun.* **2019**, *5*, 24. [CrossRef]

66. McMullin, J. Cancer and the Comics: Graphic Narratives and Biolegitimate Lives. *Med. Anthropol. Q.* **2016**, *30*, 149–167. [CrossRef] [PubMed]

67. Hamdy, S.; Nye, C.; Bao, S.; Brewer, C.; Parenteau, M. *Lissa: A Story About Medical Promise, Friendship, and Revolution*; University of Toronto Press: North York, ON, Canada, 2017.

68. El Refaie, E. *Visual Metaphor and Embodiment in Graphic Illness Narratives*; Oxford University Press: New York, NY, USA, 2019; ISBN 978-0190678173.

69. Roselfield, I.; Ziff, E.; Van Loon, B. *DNA: A Graphic Guide to the Molecule that Shock the World*; Columbia University Press: New York, NY, USA, 2011; ISBN 978-0231142717.

70. Schultz, M.; Cannon, Z.; Cannon, K. *The Stuff of Life: A Graphic Guide to Genetics and DNA*; Hill and Wang: New York, NY, USA, 2009; ISBN 978-0809089475.

71. Takemura, M.; Kikuyaro, S.O. *The Manga Guide to Biochemistry*; No Starch Press: San Francisco, CA, USA, 2011; ISBN 978-1593272760.

72. Filipe, A.; Renedo, A.; Marston, C. The co-production of what? Knowledge, values, and social relations in health care. *PLoS Biol.* **2017**, *15*, e2001403. [CrossRef]

73. Weitkamp, E.; Featherstone, H. Often overlooked: Formative evaluation in the development of Science Comics. *J. Sci. Commun.* **2009**, *8*, A04. [CrossRef]

74. Dobbins, S. Comics in public health: The sociocultural and cognitive influence of narrative on health behaviours. *J. Graph. Nov. Comics* **2016**, *7*, 35–52. [CrossRef]

75. Willis, L.A.; Kachur, R.; Castellanos, T.J.; Nichols, K.; Mendoza, M.C.B.; Gaul, Z.J.; Spikes, P.; Gamayo, A.C.; Durham, M.D.; LaPlace, L.; et al. Developing a Motion Comic for HIV/STD Prevention for Young People Ages 15–24, Part 2: Evaluation of a Pilot Intervention. *Health Commun.* **2018**, *33*, 229–237. [CrossRef] [PubMed]

76. Willis, L.A.; Kachur, R.; Castellanos, T.J.; Spikes, P.; Gaul, Z.J.; Gamayo, A.C.; Durham, M.; Jones, S.; Nichols, K.; Han Barthelemy, S.; et al. Developing a Motion Comic for HIV/STD Prevention for Young People Ages 15–24, Part 1: Listening to Your Target Audience. *Health Commun.* **2018**, *33*, 212–221. [CrossRef]

77. Alemany-Pagès, M.; Azul, A.M.; Ramalho-Santos, J. The use of comics to promote health awareness: A template using non-alcoholic fatty liver disease. *Eur. J. Clin. Investig.* **2022**, *52*, e13642. [CrossRef] [PubMed]

78. Toroyan, T.; Reddy, P.S. Participation of South african youth in the design and development of AIDS photocomics. *Int. Q. Community Health Educ.* **1997**, *17*, 131–146. [CrossRef] [PubMed]

79. Leung, M.M.; Green, M.C.; Cai, J.; Gaba, A.; Tate, D.; Ammerman, A. Fight for Your Right to Fruit: Development of a Manga Comic Promoting Fruit Consumption in Youth. *Open Nutr. J.* **2015**, *9*, 82–90. [CrossRef]

80. Kiragu, K.; Obwaka, E.; Odallo, D.; Van Hulzen, C. Communicating about sex: Adolescents and parents in Kenya. *AIDS/STD Health Promot. Exch.* **1996**, *3*, 11–13.

81. Schneider, E.F. Quantifying and Visualizing the History of Public Health Comics. In *iConference 2014 Proceedings*; iSchools: Grandville, MI, USA, 2014; pp. 995–997. [CrossRef]

82. Alemany-Pagès, M.; Ramalho-Santos, J.; Azul, A.M.; Tavares, R. *A Healthy Liver Will Always Deliver*; Coimbra University Press: Coimbra, Portugal, 2020; ISBN 978-9892620428.

83. Alemany-Pagès, M.; Moura-Ramos, M.; Araújo, S.; Macedo, M.P.; Ribeiro, R.T.; Ramalho-Santos, J.; Azul, A.M. Insights from qualitative research on NAFLD awareness with a cohort of T2DM patients: Time to go public with insulin resistance? *BMC Public Health* **2020**, *20*, 1142. [CrossRef] [PubMed]

84. Fox, S.I.; Rompolski, K. *Human Physiology*, 15th ed.; McGraw-Hill Education: New York, NY, USA, 2018; ISBN 978-1260092844.

85. Jensen, T.; Abdelmalek, M.F.; Sullivan, S.; Nadeau, K.J.; Green, M.; Roncal, C.; Nakagawa, T.; Kuwabara, M.; Sato, Y.; Kang, D.H.; et al. Fructose and sugar: A major mediator of non-alcoholic fatty liver disease. *J. Hepatol.* **2018**, *68*, 1063–1075. [CrossRef] [PubMed]

86. Lim, J.S.; Mietus-Snyder, M.; Valente, A.; Schwarz, J.-M.; Lustig, R.H. The role of fructose in the pathogenesis of NAFLD and the metabolic syndrome. *Nat. Rev. Gastroenterol. Hepatol.* **2010**, *7*, 251–264. [CrossRef] [PubMed]

87. Petersen, M.C.; Vatner, D.F.; Shulman, G.I. Regulation of hepatic glucose metabolism in health and disease. *Nat. Rev. Endocrinol.* **2017**, *13*, 572–587. [CrossRef]

88. Brunt, E.M.; Wong, V.W.-S.; Nobili, V.; Day, C.P.; Sookoian, S.; Maher, J.J.; Bugianesi, E.; Sirlin, C.B.; Neuschwander-Tetri, B.A.; Rinella, M.E. Nonalcoholic fatty liver disease. *Nat. Rev. Dis. Primers* **2015**, *1*, 15080. [CrossRef] [PubMed]

89. McNicol, S. Humanising illness: Presenting health information in educational comics. *Med. Humanit.* **2014**, *40*, 49–55. [CrossRef] [PubMed]

90. Friesen, J.; Van Stan, J., II; Elleuche, S. Communicating science through comics: A method. *Publications* **2018**, *6*, 38. [CrossRef]

91. Kuttner, P.; Weaver-Hightower, M.B.; Sousanis, N. Comics-based Research: The affordances of comics for research across disciplines. *Qual. Res.* **2021**, *21*, 195–214. [CrossRef]

92. Farinella, M. Of Microscopes and Metaphors: Visual Analogy as a Scientific Tool. *Comics Grid J. Comics Scholarsh.* **2018**, *18*, 1–16. [CrossRef]

Review

Biological Activity of Gold Compounds against Viruses and Parasitosis: A Systematic Review

Custódia Fonseca [1,2,*] and Manuel Aureliano [1,2]

1   Faculty of Science and Technology, Universidade do Algarve, 8005-139 Faro, Portugal; maalves@ualg.pt
2   Centro de Ciências do Mar (CCMar), Universidade do Algarve, 8005-139 Faro, Portugal
*   Correspondence: cfonseca@ualg.pt

Abstract: In this contribution, we provide an overview of gold compound applications against viruses or parasites during recent years. The special properties of gold have been the subject of intense investigation in recent years, which has led to the development of its chemistry with the synthesis of new compounds and the study of its applicability in various areas such as catalysis, materials, nanotechnology and medicine. Herein, thirteen gold articles with applications in several viruses, such as hepatitis C virus (HCV), influenza A virus (H1N1), vesicular stomatitis virus (VSV), coronavirus (SARS-CoV and SARS-CoV-2), Dengue virus, and several parasites such as *Plasmodium* sp., *Leishmania* sp., *Tripanossoma* sp., *Brugia* sp., *Schistosoma* sp., *Onchocerca* sp., *Acanthamoeba* sp., and *Trichomonas* sp. are described. Gold compounds with anti-viral activity include gold nanoparticles with the ligands mercaptoundecanosulfonate, 1-octanethiol and aldoses and gold complexes with phosphine and carbene ligands. All of the gold compounds with anti-parasitic activity reported are gold complexes of the carbene type. Auranofin is a gold drug already used against rheumatoid arthritis, and it has also been tested against virus and parasites.

Keywords: gold nanoparticles; gold complexes; auranofin; biological targets; virus; parasites

Citation: Fonseca, C.; Aureliano, M. Biological Activity of Gold Compounds against Viruses and Parasitosis: A Systematic Review. *BioChem* **2022**, *2*, 145–159. https://doi.org/10.3390/biochem2020010

Academic Editor: Buyong Ma

Received: 31 December 2021
Accepted: 12 May 2022
Published: 14 May 2022

Publisher's Note: MDPI stays neutral with regard to jurisdictional claims in published maps and institutional affiliations.

1. Introduction

Through the ages in most major civilizations, gold has been the chemical element that has attracted the most attention and desire, due to its unique characteristics such as its bright yellow color, corrosion resistance and extraordinary physical properties. In the form of amulets and medallions, it was used to ward off disease and evil spirits. In many cases, potions containing gold powders were administered to ill patients. After the alchemists learned to use aqua regia to dissolve gold, compounds as well as elemental gold were used in medicinal treatments [1].

Nowadays, the properties of gold are interpreted with the application of the theory of relativity, hence providing an explanation for a series of properties for the gold complexes that differ from those of its congeners, such as a preference for linear coordination, a higher stability of higher oxidation states, the formation of gold compounds and the establishment of gold–gold interactions with a similar strength to hydrogen bonds [2,3]. All these facts promoted the research in gold chemistry in different areas from nanotechnology and material chemistry [4–6] to catalysis [7–15] and medicine [16–19].

Gold compounds are well-known for their biological and medicinal applications, for instance, in rheumatoid arthritis, antibacterial, antivirus, and anti-parasite activity, as well as in Alzheimer's disease [18,20–26]. The mechanisms of action of gold compounds are mostly unknown; however, several proteins have been studied as possible targets, such as tyrosine phosphatases, aquaporins and P-ATPases, among others [17,26–29]. However, the putative pharmacological targets of gold compounds are not the objective of this review, and the reader should look elsewhere for this topic [17,18,22,27,29]. Herein, we aim to reviewed recent applications of gold compounds against virus and parasites. It was observed that, after a search carried out in the Web of Science with the keyword "gold compound", 99,014 articles were published in the last 10 years (2012–2021), more than triple the total in the previous period (2002–2011). Another fact to bear in mind is the steady increase over time, especially in the field of medicine. Although there are many articles and studies that overlap between the field of medicine and nanotechnology, specific papers on gold compounds in these categories, plus catalysis, can be found (Figure 1).

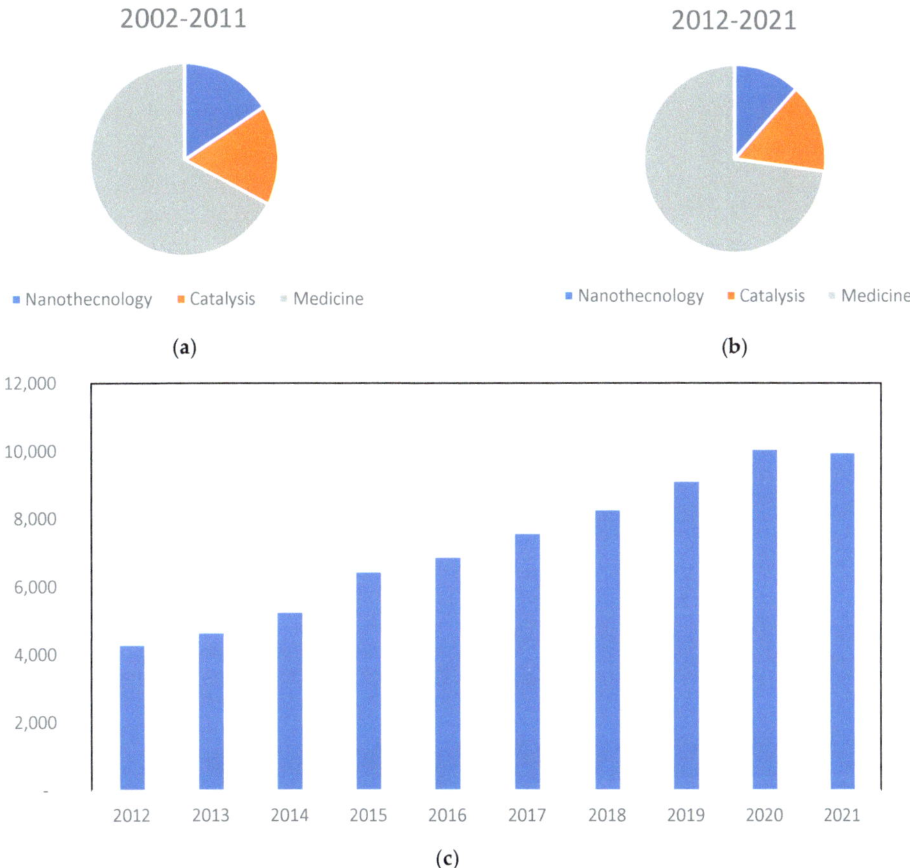

Figure 1. Number of articles indexed on the Web of Science obtained using the keywords "gold compounds" and "nanotechnology", "gold compounds" and "catalysis" and "gold compounds" and "medicine". (**a**) Graph referring to the decade of 2002–2011; (**b**) graph referring to the decade of 2012–2021; (**c**) graph referring to articles published using gold compounds in medicine from 2012 until now. Review articles were excluded.

Although the first widespread application of gold compounds was against pulmonary tuberculosis, for which they do not provide an effective therapy, it is in the treatment of rheumatoid arthritis that gold compounds have proved effective [30]. Auranofin, triethylphosphine (2,3,4,6-tetra-*O*-acetyl-*β*-1-D-thiopyranosato-*S*) gold(I) is the example of this, as it continues to be used in the clinic today and is the subject of research [31,32]. Other gold compounds are also under investigation in the treatment of diseases such as cancer, viruses such as human HIV and SARS-CoV-2 and parasitosis such as malaria and amoebiasis (Figure 2).

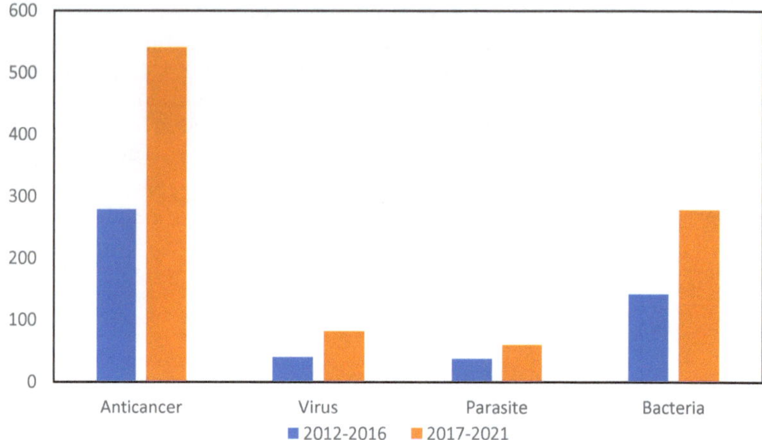

Figure 2. Number of articles indexed to Web of Science where gold compounds are investigated for biological activity against cancer, parasitosis and viruses in last decade (2012–2021).

This paper will present a systematic review on the topic of "biological activity of gold compounds against virus and parasites". The information obtained from the database is grouped and analyzed to examine the applications of gold molecules in diseases caused by those two types of organisms.

2. Methodology

In this systematic review, the Web of Science was chosen as a database. The search terms were "gold compound" and the target "virus" or "parasite". The literature search was carried out for the last two years (2020–2021).

The initial research was carried out by introducing the keywords in the database, Web of Science, and the range date was 1 January 2020 to 30 December 2021. From the set of articles obtained, review articles were excluded and then the titles and abstracts were read in order to check if the study reported in the articles was within the theme of this review article, then it could be included or excluded (Figure 3).

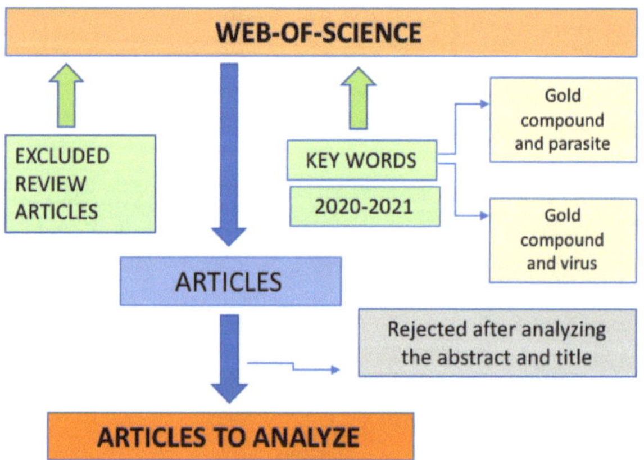

Figure 3. Diagram of the method used to select the articles to be analyzed.

The analysis of the articles considered the type of compound studied and where the biological activity was determined.

3. Gold Compounds with Biological Activity against Virus and Parasites

3.1. Gold Compounds Used in Anti-Virus Studies

From the research carried out in Web of Science with the words "gold compound" and "virus" for the years 2020–2021, 43 articles were obtained, and review articles were excluded, leaving 30. These articles were analyzed by reading the title and abstract, leaving six articles under analysis, which form the basis of this review.

Molecular gold(I) species were examined for the inhibition pf liver fibrosis and the hepatitis C virus (HCV). Four gold(I) complexes, auranofin, sodium aurothiomalate, Ph_3PAuCl and (PTA)AuCl (Table 1), were tested against HSC line Lx2, human hepatoma cells and Huh—K2040 cells. Two of these molecules, Ph_3PAuCl [24,33] and (PTA)AuCl [34], were synthetized according to the references.

Table 1. Gold compounds with structures and biological effects against viruses and parasitosis.

Viruses/Parasitosis	Gold Compound/Structure	Ref
Fibrosis and hepatitis C virus (HCV infection)	*Gold complexes:* auranofin; sodium aurothiomalate; Ph_3PAuCl; (PTA)AuCl; 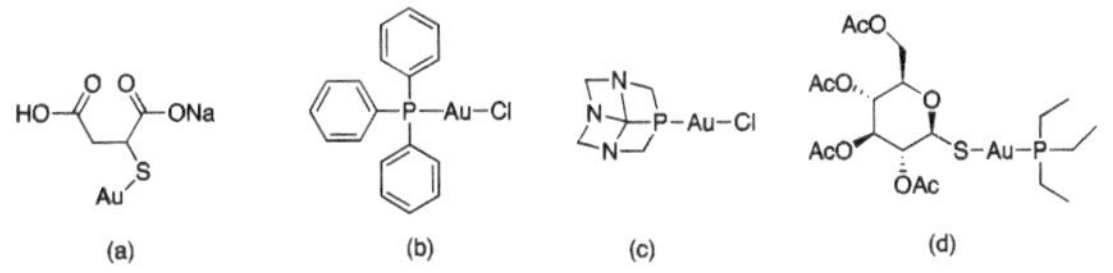 Gold(I) compounds: (a) sodium aurothiomalato; (b) Ph_3PAuCl; (c) (PTA)AuCl; (d) auranofin	[24]
H1N1	*Nanoparticle:* gold (Au) core and the ligands mercaptoundecansulfonate (MUS) and 1-octanethiol (OT) 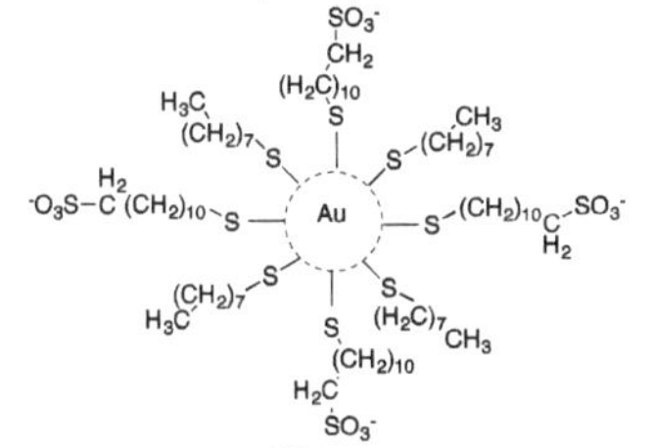 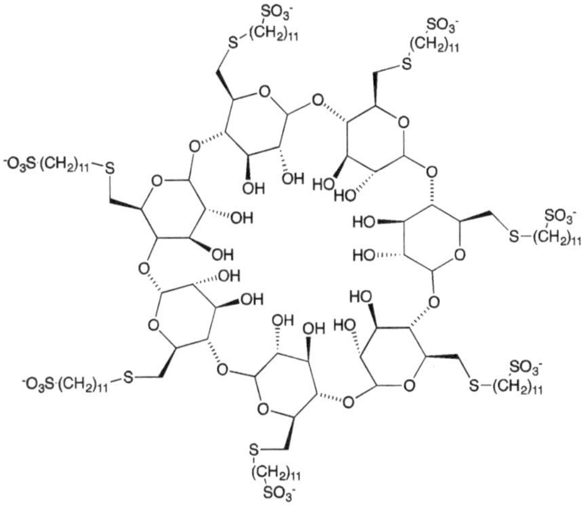	[35]

Table 1. *Cont.*

Viruses/Parasitosis	Gold Compound/Structure	Ref
H1N1	*Nanoparticles*: gold glyconanoparticles 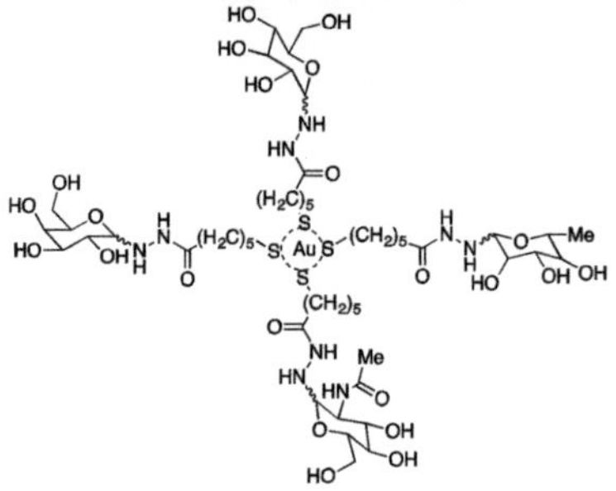	[36,37]
SARS-CoV-2	Auranofin 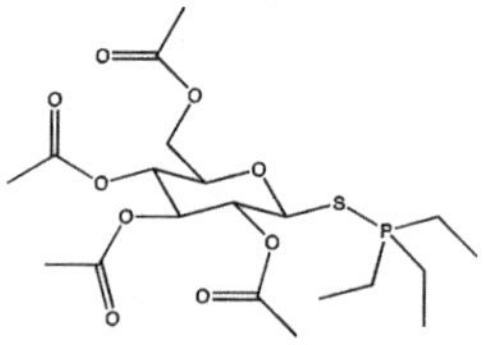	[38]
Dengue virus	*Nanoparticles:* gold nanoparticles coated with ω-terminated with sugars bearing multiple sulfonate groups (L-AuNPs) 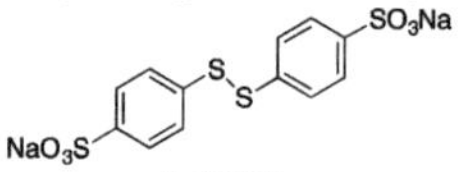 L-AuNPs	[39]
SARS-CoV-2	*Gold(I) complexes* with *N*-heterocyclic carbene (NHC) or gold(III)–dithiocarbamato complexes 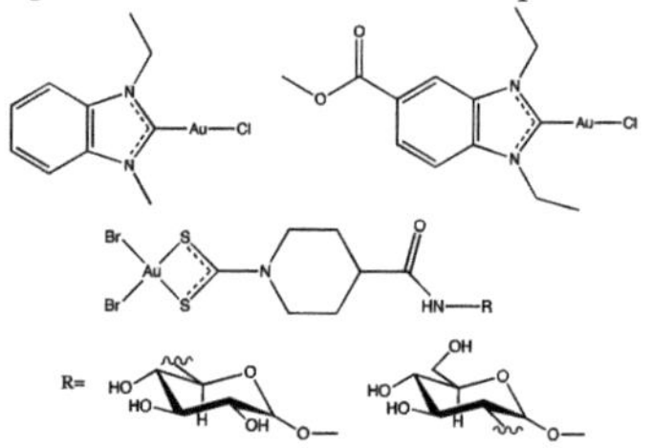	[40]
Plasmonium falciparum and *Leishmania infantum*	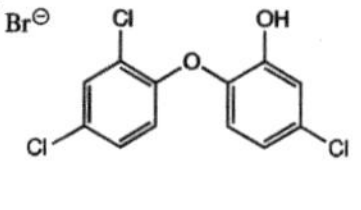 *Gold(I) complexes* combined with triclosan 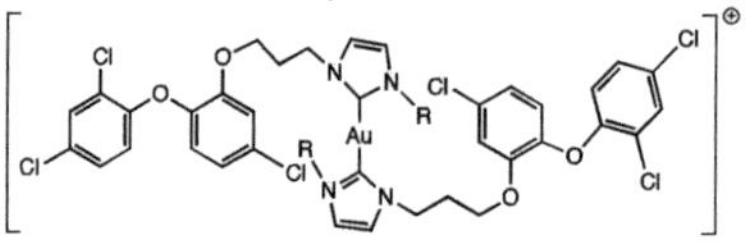 a) R=Me; b) R = Mes; c) R=Bn; d) R= Quin; e) R = C$_6$H$_4$SMe Triclosan	[41,42]

Table 1. *Cont.*

Viruses/Parasitosis	Gold Compound/Structure	Ref
Protozoa and Helminth infections	*Gold(I) com-* *plex* 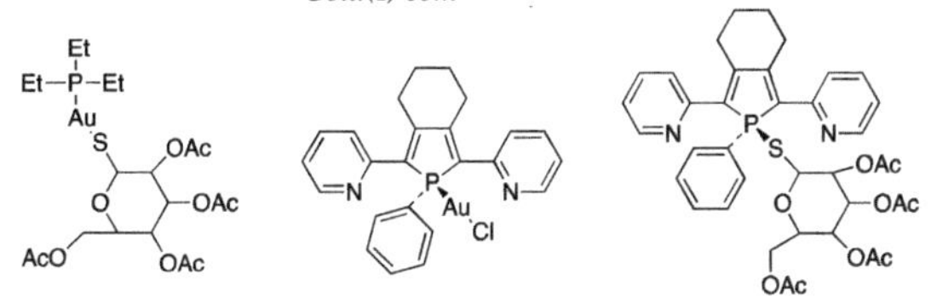 Auranofin GoPI GoPI-sugar	[43]
Leishmania major, Toxoplasma gondii and Trypanosoma brucei parasites	Gold(I)–carbene complexes 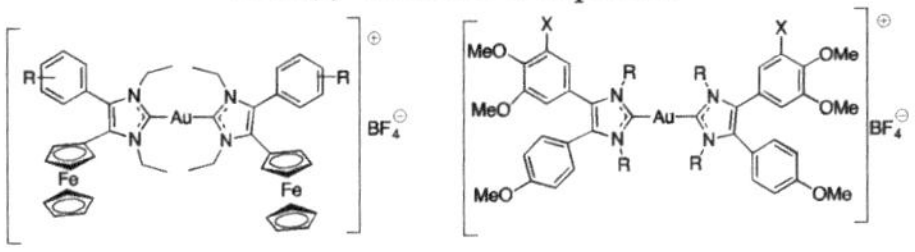 a) R= H; b) R= 3,4,5-(OMe)₃; c) R= 4-OMe d) = Me; X= OMe; e) R= Et, X= OMe f) R= Me, X= Cl; g) R= Et, = Br	[44]
Trichomonas vaginalis	*Gold Complexes:* gold(I) phosphine derivatives 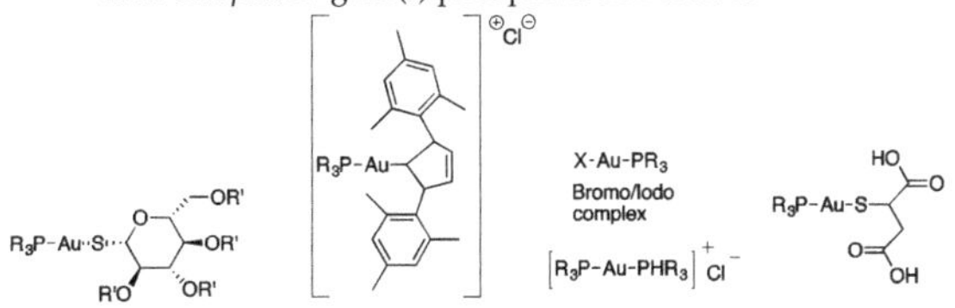 Thioglucose complex N-Heterocyclic carbene complex Bis-phosphine complex Thiomalic acid complex	[45]
Helmintic	Gold(I) nanoparticles 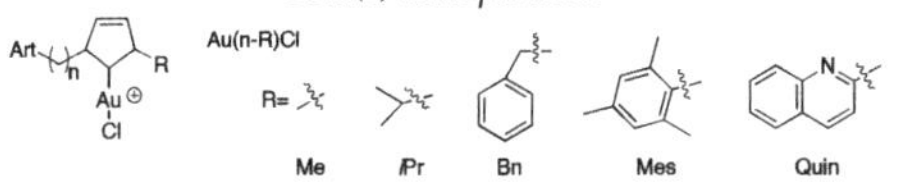 R= Me, ʲPr, Bn, Mes, Quin	[46]
L. amazenensis and L. braziliensis	*Gold(I) and gold(III) complexes* 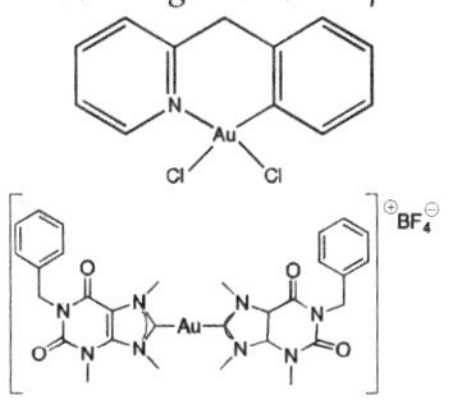	[47]

As a preliminary screen for anti-fibrotic activity, we used the Lx-2 cell morphology. Auranofin and Ph_3PAuCl with 2 µM and 5 µM concentrations were the only active compounds, as they revert back to a quiescent state of the cells.

After being treated with auranofin (2 µM and 5 µM) and Ph_3PAuCl (2 µM and 5 µM), the HSC were lysed and type I collagen expression and the proteins STAT3 and SMAD2 were determined. As compared to DMSO-treated human hepatic stellate cells (HSCs), auronafin treatment with concentrations of 2 µM and 5 µM resulted in 50% and 40% decreases in type I collagen expression, respectively. Ph_3PAuCl also reduced type I collagen expression, but less dramatically, with only a 10% and 20% reduction at 2 µM and 5 µM, respectively. Next, two proteins, STAT3 and SMAD2, which play roles in type I collagen were determined. At both 2 µM and 5 µM concentrations, auranofin showed a complete abrogation of STAT3 phosphorylation, and also a dramatic increase in SMAD phosphorylation.

Ph$_3$PAuCl showed a 60% and 90% inhibition of STAT3 phosphorylation at concentrations of 2 µM and 5 µM, respectively, and also resulted in an increase in SMAD2 phosphorylation, albeit to a lesser extent than auranofin. None of the gold compounds appeared to have a significant impact on α-SMA protein levels. The auranofin 2 µM and 5 µM concentrations show an 80% and 98% decrease, respectively, in type I collagen mRNA, while the α-SMA mRNA levels decreased 80% and 96%, respectively. Meanwhile, for Ph$_3$PAuCl, the decrease in collagen-I mRNA expression is not noticeable, while α-SMA expression is nearly 70% and 50% lower, respectively. A decrease in NS3 and NS5a expression in Huh7 K2040 cells was observed after being treated with auranofin and Ph$_3$PAuCl. Auranofin at 5 µM inhibits 60% of HCV-NS5a, while it causes 50% inhibition of HCV-NS3 and a complete abrogation of STAT3 phosphorylation. In conclusion, these results indicate the inhibitory potential of these drugs against HCV protein expression [24].

Cagno, V. et al. designed two compounds: in the first compound named MUST: OT NP (Table 1), the moiety was fixed on gold nanoparticles (4 nm gold nanoparticles (NPs) coated with a mixture of octane thiol (OT) and mercapto-undecane sulfonic acid (MUST) [35,48,49]); in the second compound, named CD1, the moiety was linked to the primary face of β-cyclodextrins (CD) [50]. In previous study, it was demonstrated that gold nanoparticles coated with mercapto-undecane sulfonic acid (MUS) inhibit heparan sulfate proteoglycan (HSPG)-dependent viruses irreversibly while retaining the low-toxicity profile [51]. These compounds display inhibitory activity in the absence of toxicity, with 50% effective concentrations, (EC$_{50}$), values between 1.38 and 12.0 µM/m and for H1N1 stains. For vesicular stomatitis virus (VSV), the EC$_{50}$ value is 0.053 µM/mL relative to compound MUS:OT-NP; for CD1, the EC$_{50}$ determined was higher (for VSV, 45 µM, and for H1N1, between 6.28 and 53.2 µM).

The antiviral action is irreversible for influenza A virus (H1N1), while for VSV, the inhibition is reversible. These results further broaden the spectrum of activity of MUS-coated gold nanoparticles [35] (Table 1).

Synthesized gold glyconanoparticles [37] (Table 1) were tested against the *Influenza* virus H1N1 (A/Puerto Rico/8/34) via a standard protocol [36]. The nanoparticles revealed low cytotoxicity towards the MDCK cells (higher than 100 µg/mL) and high antiviral activity at concentrations of 3 and 6 µg/mL. The strongest antiviral activity was observed for the sample with a glycoligand composition GlcNAc:Man:Fuc = 20:75:5% mol.

These preliminary data on the activity against the A/Puerto Rico/8/34 (H1N1) demonstrated the prospects of their further investigation for the search for efficient drugs for the prevention and treatment of acute respiratory viral infections [36] (Table 1).

Among more than 100 structurally diverse metal complexes, 36 gold(I) and (III) complexes with different types of co-ligands, including *N*-heterocyclic carbenes (NHCs), alkynyls, dithiocarbamates, phosphines and chlorides, were selected for profiling as inhibitors of SARS-CoV-2. The inhibition can occur by means of one of two replication mechanisms, namely the interaction of the spike (S) protein with the ACE2 receptor and the papain-like protease PLpro. The chloroauric acid (HAuCl$_4$; oxidation state + 3) has moderate inhibition (about 47% inhibitory activity, the other gold compounds were poorly active or inactive) [40].

Based on previous results, the chosen inhibitor concentration was 10 µM for assays with SARS-CoV PLpro, while 1.0 µM was used for SARS-CoV-2 PLpro. Ten gold complexes were tested against PLpro of SARS-CoV, and the values determined range from 0.3 to 1.2 µM, and against SARS-CoV-2 PLpro, the values determined range from 0.1 to 1.5 µM. These results confirm the high potential of gold compounds as protease-inhibiting antiviral drugs. For gold(III) dithiocarbamates, compounds 3 and 4 (Figure 4) had an IC$_{50}$ of 0.21 µM and 0.09 µM, respectively. These molecules were the most active SARS-CoV-2 PLpro inhibitors identified in this study and show strong preference for the enzyme of SARS-CoV-2. Compounds 1 and 2 (Figure 4), with an IC$_{50}$ of 0.35 µM and 0.33 µM, were stronger inhibitors for the SARS-CoV enzyme.

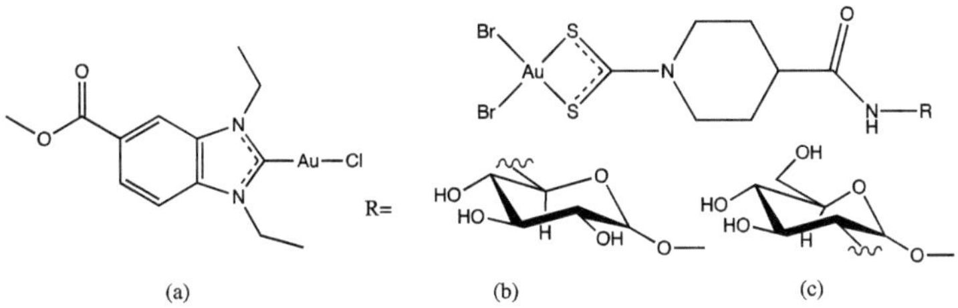

Figure 4. Gold complexes tested against SARS-CoV-2: (**a**) Gold(I) NHC complex; (**b**) and (**c**) are Gold(III)-dithiocarbamato complexes.

The evaluation of structure–activity relationships indicated a preference for complexes with good leaving groups (e.g. chloride) over compounds with firmly coordinated ligands such as dicarbene gold complexes of the type $(NHC)_2Au^+$, which were inactive. Exceptionally strong and selective activity against SARS-CoV-2 PLpro was obtained with the gold(III)-dithiocarbamato glycoconjugates, compounds (b) and (c) in Figure 4. Most of the gold complexes tested showed too strong toxicity; however, for four compounds, very low or missing toxicity against the Caco-2 cell line was noted, and these four complexes were selected for the SARS-CoV-2 antiviral assays. Among these, the gold(I) NHC complex (a) showed very promising activity at low micromolar concentration, and two gold(III)-dithiocarbamato complexes (b) and (c), Figure 4,were strongly active at the highest applied nontoxic dosage [40].

Auranofin inhibits SARS-CoV-2 replication in human cells (Huh7 cells) at a low micromolar concentration (EC_{50} 1.4 μM). The treatment of cells with auranofin resulted in a 95% reduction in the viral RNA at 48 h after infection. Auranofin treatment dramatically reduced the expression of SARS-CoV-2-induced cytokines in human cells. These data indicate that auranofin could be a useful drug to limit SARS-CoV-2 infection and associated lung injury due to its antiviral, anti-inflammatory and anti-reactive oxygen species (ROS) properties. Further animal studies are warranted to evaluate the safety and efficacy of auranofin for the management of SARS-CoV-2 associated disease [38].

Different gold nanoparticles (AuNP) coated with ligands ω-terminated with sugars bearing multiple sulfonate groups (L-AuNPs) (Table 1) were tested against Dengue virus (DENV). The ligands varied in length, in number of sulfonated groups, as well as their spatial orientation induced by the sugar head groups. Two candidates were identified, a glucose- and a lactose-based ligand showing a low EC_{50} (effective concentration that inhibits 50% of the viral activity) for DENV-2 inhibition, moderate toxicity and a virucidal effect in hepatocytes with a titer reduction in the median tissue culture infectious dose $log_{10}TCI\ D_{50}$ 2.5 and 3.1. Molecular docking simulations complemented the experimental findings, suggesting a molecular rationale behind the binding between sulfonated head groups and DENV-2 envelope protein [39].

3.2. Gold Compounds Used in Anti-Parasite Studies

From the research carried out in Web of Science with the words "gold compound" and "parasite" for the years 2020–2021, 27 articles were obtained, and review articles were excluded, leaving 23. These articles were analyzed by reading the title and abstract, leaving six articles under analysis, which form the basis of this review.

Ouji et al. synthesized and characterized five hybrid molecules combining triclosan and gold(I) complexes, gold(I)-NHC-TC (Table 1), and evaluated them, in vitro, to determine the activities on *Plasmodium falciparum* and *Leishmania infantum* parasites [41,42].

The antimalarial activity of the hybrid molecules was tested using *P. falciparum* strain F32-TEM and the IC_{50} determined range from 120 nM to 1465 nM. These values were determined in comparison with the corresponding proligand, triclosan, and artemether, the antiplasmodial control drug. Globally, the presence of gold(I) can have opposite effects, with an improvement in efficacy compared to the corresponding proligand; triclosan showed weak antiplasmodial activity with an IC_{50} value of 6000 nM, and the artemether had an IC_{50} of 6.1 nM.

All the hybrid compounds and their carbene precursors were also screened in vitro on both the promastigote and axenic amastigote stages of *L. infantum*, and the IC_{50} values compare to reference drugs for Leishmaniosis, as well as triclosan and auranofin. The results obtained against promastigotes of *L. infantum* show moderate or no activity with an IC_{50} range from 5.51 μM to 63.87 μM. Looking at the activities against promastigotes of the proligand/complex couples, there was just one clear enhanced addition of gold(I) in the compound a (Figure 5). In the case of axenic amastigotes *L. infantum*, the IC_{50} was 0.21–1.54 μM. The data observed could not be totally established in relation to the amastigote activity. These Au(I)bis(NHC-TC) series had similar activity to previously reported cationic Au(I)bis(NHC) and neutral Au(I)(NHC)Cl complexes including aliphatic and aromatic substituents [52,53], which suggests that no enhanced activity was found by the addition of the triclosan moiety. For compound a (Figure 5), a selectivity index (S.I.) value of 7.5 was determined, which was comparable to 6.0 for triclosan; these low S.I. values mean that compound a has the same range of activity against normal mammalian cells and *P. falciparum*. Concerning Leishmania, the S.I. value of amastigote/promastigote ranged from 4.99 to 14.81 for the proligands, and from 5.50 to 26.65 for the complexes, with increased selectivity in most cases by the coordination of the gold(I) on the NHCs. Auranofin exhibited higher activity and selectivity (IC_{50}(promastigotes) 4.32 μM; S.I. 2.67 and IC_{50} (amastigotes) 0.07 μM; S.I. 38.14) and on two stages of *L infantum* than all complexes. The results obtained demonstrate that the incorporation of the triclosan derivative in complexes could be related to a loss in selectivity, without increasing the activity.

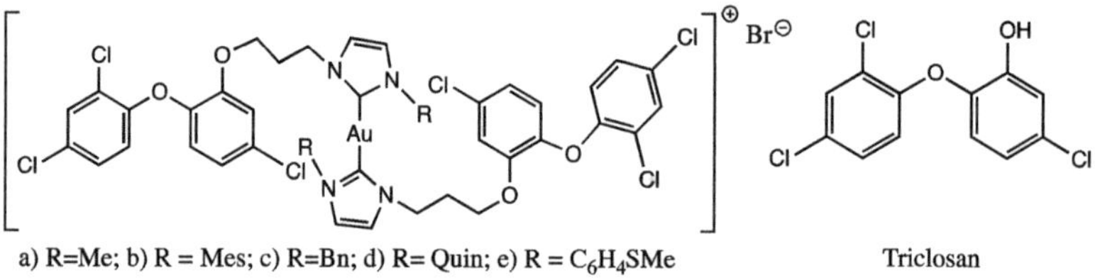

Figure 5. Gold(I)-NHC-TC hybrid molecule and triclosan compound screening against *Leishmania* sp. and *Plasmodium* sp.

However, the cytotoxicity assays suggest a structure modulation to improve the selectivity and the IC_{50} values too.

The synthesized molecules were also tested in the context of *Plasmodium falciparum* artemisinin-resistance (strain F32-ART), and the results show no cross-resistance induced between one gold(I) complex and artemisinin [41].

The same author as previously reported synthesized and fully characterized three families of gold(I) NHC complexes which incorporated a covalently attached dihydroartemisinin (DHA) derivative (Figure 6). The proligands (Ln-R) and the complexes (Au-*bis*(n-R) and Au(n-R)Cl), as well as reference molecules (artemisinin, artemether, auranofin and the proligand precursor DHA-C3), were screened in vitro against the *P. falciparum*, and the cytotoxicity on *Vero* cells was evaluated to determine the selectivity of the most active compounds. Among the 29 complexes compounds tested, an IC_{50} value range from 9 to 104 nM was obtained, and ten of them showed high antiplasmodial activities, with IC_{50} values less than 50 nM. Auranofin, used as a gold reference molecule, was not active against *P. falciparum* parasites, with a higher IC_{50} value of 1.5 μM. It was possible to establish a structure–activity relationship, and thus the potency of the complexes containing methyl or benzyl groups on the NHCs increased with the length of the spacer, whereas no correlation was spotted for the mesityl and the quinoline series. The selectivity indexes for the gold(I) complexes were between 8 and 178, with molecules with aliphatic R groups (Me, iPr) having the best selectivity with S.I. values of 143 and 178.

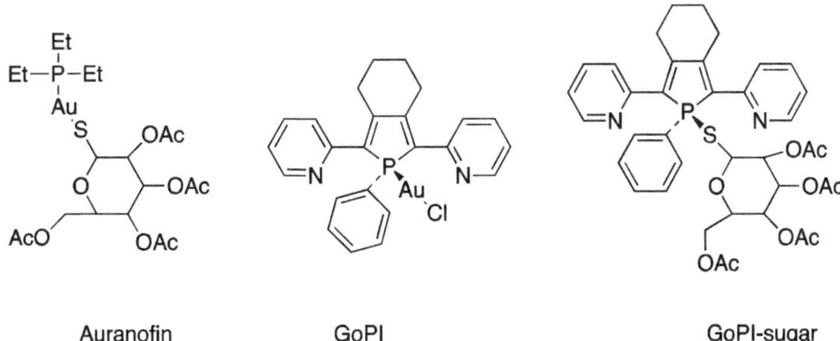

Figure 6. Dihydroartemisinin derivatives and gold(I)-*N*-heterocyclic carbene (NHC) complexes.

Three hybrid molecules with the best selectivity indexes Au-*bis*(3-Me), Au-*bis*(4-Me) and Au-*bis*(5-Me) were evaluated in vitro for their efficacy in the context of resistance to artemisinins. The results obtained confirm the existence of a cross-resistance between artemisinin and the hybrid molecules, which can be explained by the DHA part of the hybrid, responsible for the quiescence entrance of the parasites and the lack of activity of the NHC part at the mitochondrial level due to limited access or pharmacodynamic properties. These data are in accordance with previously obtained results which highlighted the risks of parasite cross-resistance between artemisinins and endoperoxide-based compounds [42,52,54].

The screening of various parasites—protozoans, trematodes, and nematodes—was undertaken to determine the in vitro killing activity of a new gold(I) complexes, GoPI-sugar (Figure 7). GoPI-sugar is a novel 1-thio-β-D -glucopyranose 2,3,4,6-tetraacetato-*S* -derivative that is a chimera of the structures of GoPI and auranofin, designed to improve the stability and bioavailability of GoPI (Figure 7). GoPI-sugar was found to efficiently kill intra-macrophagic *Leishmania donovani* amastigotes and adult filarial and trematode worms. Among new gold(I) complexes, the phosphole-containing gold(I)complex {1-phenyl-2,5-di(2-pyridyl)phosphole}AuCl (abbreviated as GoPI) is an irreversible inhibitor of both purified human glutathione and thioredoxin reductases. Table 2 includes the results of biological activity of the GoPI-sugar compound against different parasites, expressed in killing % activity, motibility % inhibition and 50% inhibition concentration (IC_{50}) [43].

Figure 7. Structures of gold(I) complexes screening against various parasites.

Miyamoto et al. synthesized several gold(I) phosphine complexes: nonacetylated and paracetylated thioglucose-Au-PR_3 complexes; thiomalic acid-Au-PR_3 complexes; NHC-Au-PR_3 complexes; PR_3-Au-Cl complexes; [PR_3-Au-PR_3]Cl; and PR_3-Au(I)complexes (Figure 8) to study their activity against *Trichomonas vaginalis*. A systematic structure–activity relationship demonstrates that the diversification of gold(I) complexes, particularly as halides with simple trialkyl phosphines of the C1−C3 chain length or as *bis*(trialkyl phosphine) complexes, can markedly improve potency against *T. vaginalis* and selectivity. All gold(I) complexes effectively inhibited the two most abundant isoforms of the presumed target enzyme, TrxR, but a subset of compounds were markedly more active against live *T. vaginalis* than TrxR, suggesting that alternative targets exist. All of the tested gold(I) complexes inhibited TrxR in *T. vaginalis*, with the most potent compounds exhibiting IC_{50} values of <50 nM, little difference in the inhibition of two of the most abundantly expressed TrxR isoforms, and good correlations between the inhibition of individual isoforms and the inhibition of total cellular TrxR activity. Two of the compounds with increased activity against whole cells compared to TrxR were gold(I)-bis-phosphines [45].

Table 2. Biological activity (worm killing activity; motility inhibition; in vitro activity) of a gold complex, GoPI-sugar, in different species [43].

Parasite	Biological Activity of GoPI-Sugar [a]
Schistosoma mansoni	[GoPI-sugar] = 5 µM 100% dead in day 1 [GoPI-sugar] = 2.5 µM 100% dead in day 5
Brugia pahangi (worms)	[GoPI-sugar] = 10 µM 100% inhibition in day 3
Brugia pahangi (worms)	IC_{50} 1.7 µM day 6
Onchocerca ochengi (adults)	100% inhibition motility day 5
Tripanossoma b. gambiense (trypomastigotes)	IC_{50} 1.11 µM
Tripanossoma b. brucei (trypomastigotes)	IC_{50} 1.83 µM
Tripanossoma cruzi (amastigotes)	IC_{50} 0.56 µM
Leishmania infantum (amastigotes)	IC_{50} 2.38 µM
Leishmania donovani (Axenic amastigotes)	IC_{50} 1.45 µM
Leishmania donovani (intramacrophage amastigotes)	IC_{50} 0.42 µM
Acanthamoeba castellanii	IC_{50} 13.04 µM

[a] GoPI-sugar is 1-thio-β-D-glucopyranose-2,3,4,6-tetraacetato-S-dervative [55].

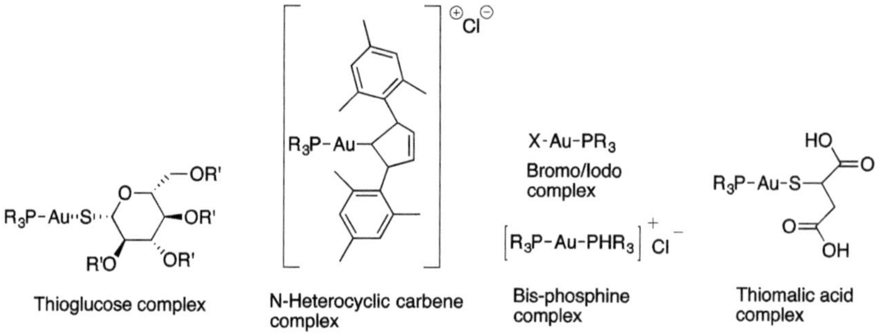

Figure 8. Gold(I) phosphine complexes.

Koko et al. synthesized a series of cationic gold(I) carbenes with various 4,5-diarylimida zolylidene as ligands (Figure 9) to test their activity against *Leishmania major*, *Toxoplasma gondii* and *Tripanossoma brucei* parasites [44].

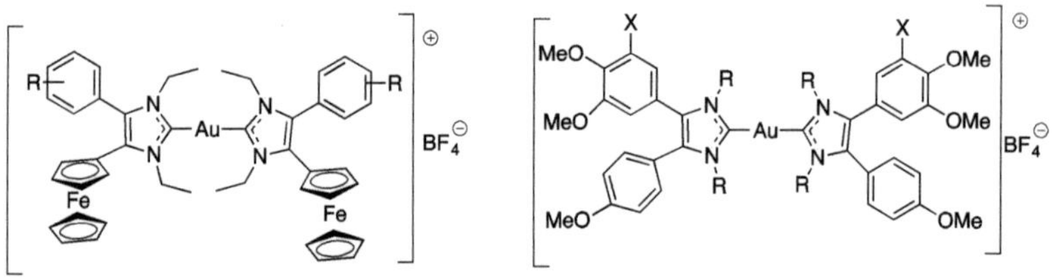

a) R= H; b) R= 3,4,5-(OMe)₃; c) R= 4-OMe d) = Me; X= OMe; e) R= Et, X= OMe f) R= Me, X= Cl; g) R= Et, = Br

Figure 9. Structures of the *N*-heterocyclic carbene-gold(I) complexes.

The results of this work are presented in Table 3. Ferrocene compound a–c (Figure 9) showed the highest activities against *L. major* amastigotes and *T. gondii* and distinct selectivity for *T. gondii* cells when compared with the activity against nonmalignant Vero cells. The ferrocene derivatives (a–c) (Figure 9) are generally more active against the *L. major* amastigotes and the *T. gondii* tachyzoites than the other tested anisyl gold complexes and the approved drugs atovaquone and amphotericin B. Compound f (Figure 9) showed the highest selectivity for *L. major* promastigotes; thus, it was the most active compound against *L. major* promastigotes of this series of compounds. The 3,4,5-trimethoxyphenyl analog 1b also exhibited a much greater selectivity for *T. b. brucei* cells when compared with its activity against human HeLa cells [44].

Table 3. Biological activity of cationic gold(I) carbene against *Leishmania major*, *Toxoplasma gondii* and *Tripanossoma brucei* parasites [44].

Parasite	Biological Activity of Cationic Gold(I)-Carbene Complexes with 4,5-Diarylimidazolylidene Ligands a
Tripanossoma gondii	EC_{50} 0.013 µM—Compound a [1] EC_{50} 0.195 µM—Compound d [1] S.I. (Vero/*T. gondii*) 28.1—Compound a [2]
Leishmania major promastigotes	EC_{50} 0.31 µM—Compound f [1] EC_{50} 3.11 µM—Compound e [1]
Leishmania major amastigotes	EC_{50} 0.11 µM—Compound a [1] EC_{50} 0.46 µM—Compound f [1] S.I. (Vero/*T. gondii* pro) 2.16—Compound d [2] S.I. (Vero/*T. gondii* amas) 3.321—Compound a [2]
Tripanossoma b. brucei	IC_{50} 0.000092 µM—Compound a [1] IC_{50} 0.028 µM—Compound c [1] S.I. (HeLa/*T. b.brucei*) 168—Compound b [2]

[1] Best and worst values of effective concentrations EC_{50} relative to tested compounds, represented in Figure 9. [2] Selectivity index (S.I. IC_{50}/EC_{50}).

Minori, K. et al. synthesized a series of gold(I) and gold(III) complexes in order to test them against *Leishmania amazonensis* and *Leishmania braziliensis*. One of the cationic Au(I) *bis*-N-heterocyclic carbenes [(c), Figure 10] has low EC_{50} values (ca. 4 µM) in promastigotes cells and no toxicity in host macrophages. Together with two other Au(III) complexes [(a) and (b), Figure 10], the compound is also extremely effective in intracellular amastigotes from *L. amazonensis*. Initial mechanistic studies include an evaluation of the gold complexes' effect on *L. amazonensis*' plasma membrane integrity [47].

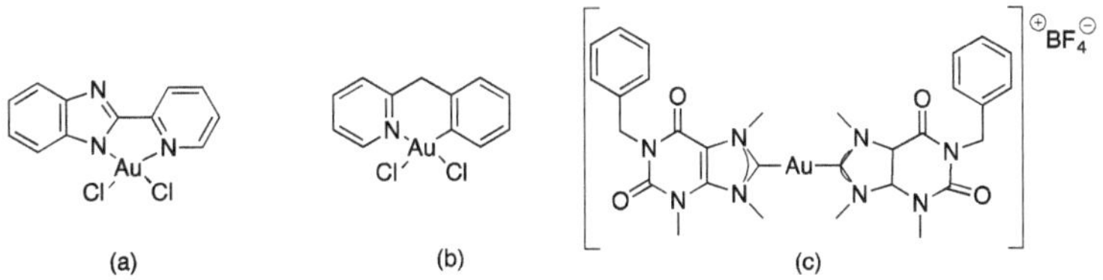

Figure 10. Structure of Au(I) and Au(III) complexes tested in biological activity against *Leishaminia* sp; (**a**) Au(III) complex; (**b**)Au(III) complex; (**c**) Cationic Au(I) bis-N-heterocyclic carbenes.

4. Conclusions

From research works presented here, most of the compounds used are Au complexes, as opposed to nanoparticles of the same metal (see Table 1).

Gold nanoparticles have been used in many applications including drug delivery, cancer diagnostics and therapy, medical imaging and non-optical biosensor [56]. The nanoparticles used in the studies presented are similar to but also different from each other; most of them have sulfur as a binding element to gold and then have aliphatic hydrocarbon chains or cyclic monomers. Gold complexes have *N*-heterocylic ligands, carbenes or phosphines, which are the most common Au complexes. Auranofin, the gold drug already used against rheumatoid arthritis, was also tested against viruses and parasites.

The mechanism of action of these compounds is almost entirely unknown, since the number of compounds used for these studies is small, not allowing the establishment of a structure–activity relationship.

Finally, this review demonstrates the diversity of structures and application of gold compounds in viral and parasites diseases. Additionally, in order for the use of gold compounds in the treatment of viruses or parasitosis to be considered, further studies are needed, since different mechanisms of action might be involved. Putting it all together, taking into consideration its role in biomedical sciences, the future of gold compounds against virus and parasites could be bright.

Author Contributions: Conceptualization, C.F. and M.A.; formal analysis C.F.; funding acquisition, M.A.; investigation, C.F. and M.A.; methodology, C.F. and M.A.; supervision, M.A.; writing—original draft, C.F.; writing—review and editing, C.F. and M.A. All authors have read and agreed to the published version of the manuscript.

Funding: This study received Portuguese national funds from FCT—Foundation for Science and Technology through the projects UIDB/04326/2020 and LA/P/0101/2020(M.A. and C.F.).

Institutional Review Board Statement: Not applicable.

Informed Consent Statement: Not applicable.

Data Availability Statement: Not applicable.

Conflicts of Interest: The authors declare no conflict of interest.

Abbreviations

HCV	hepatitis C virus
HSC	(human) hepatic stellate cells
MUST	mercapto-undecane sulfonic acid
OT	octane thiol
CD	β-cyclodextrins
ACE2	Angiotensin converting enzyme 2
S	spike
PLpro	papain-like protease
ROS	reactive oxygen species
AuNPs	gold nanoparticles
EC_{50}	Effective concentration that inhibits 50% of the viral/cell activity
DENV	Dengue virus
DHA	dihydroartemisinin
S.I.	selectivity index

References

1. Higby, G.J. Gold in Medicine: A Review of Its Use in the West before 1900. *Gold Bull.* **1982**, *15*, 130–140. [CrossRef] [PubMed]
2. Mirzadeh, N.; Privér, S.H.; Blake, A.J.; Schmidbaur, H.; Bhargava, S.K. Innovative Molecular Design Strategies in Materials Science Following the Aurophilicity Concept. *Chem. Rev.* **2020**, *120*, 7551–7591. [CrossRef] [PubMed]
3. Schmidbaur, H.; Schier, A. Aurophilic Interactions as a Subject of Current Research: An up-Date. *Chem. Soc. Rev.* **2012**, *41*, 370–412. [CrossRef] [PubMed]
4. Stratakis, M.; Garcia, H. Catalysis by Supported Gold Nanoparticles: Beyond Aerobic Oxidative Processes. *Chem. Rev.* **2012**, *112*, 4469–4506. [CrossRef]
5. Yeh, Y.-C.; Creran, B.; Rotello, V.M. Gold Nanoparticles: Preparation, Properties, and Applications in Bionanotechnology. *Nanoscale* **2012**, *4*, 1871–1880. [CrossRef]
6. Chakraborty, I.; Pradeep, T. Atomically Precise Clusters of Noble Metals: Emerging Link between Atoms and Nanoparticles. *Chem. Rev.* **2017**, *117*, 8208–8271. [CrossRef]
7. Hashmi, A.S.K. Dual Gold Catalysis. *Acc. Chem. Res.* **2014**, *47*, 864–876. [CrossRef]
8. Hashmi, A.S.K.; Hutchings, G.J. Gold Catalysis. *Angew. Chem. Int. Ed.* **2006**, *45*, 7896–7936. [CrossRef]
9. Li, Z.; Brouwer, C.; He, C. Gold-Catalyzed Organic Transformations. *Chem. Rev.* **2008**, *108*, 3239–3265. [CrossRef]
10. Arcadi, A. Alternative Synthetic Methods through New Developments in Catalysis by Gold. *Chem. Rev.* **2008**, *108*, 3266–3325. [CrossRef]

11. Corma, A.; Leyva-Pérez, A.; Sabater, M.J. Gold-Catalyzed Carbon−Heteroatom Bond-Forming Reactions. *Chem. Rev.* **2011**, *111*, 1657–1712. [CrossRef] [PubMed]

12. Gorin, D.J.; Toste, F.D. Relativistic Effects in Homogeneous Gold Catalysis. *Nature* **2007**, *446*, 395–403. [CrossRef] [PubMed]

13. Gorin, D.J.; Sherry, B.D.; Toste, F.D. Ligand Effects in Homogeneous Au Catalysis. *Chem. Rev.* **2008**, *108*, 3351–3378. [CrossRef] [PubMed]

14. Obradors, C.; Echavarren, A.M. Gold-Catalyzed Rearrangements and Beyond. *Acc. Chem. Res.* **2014**, *47*, 902–912. [CrossRef]

15. Gaillard, S.; Cazin, C.S.J.; Nolan, S.P. N-Heterocyclic Carbene Gold(I) and Copper(I) Complexes in C–H Bond Activation. *Acc. Chem. Res.* **2012**, *45*, 778–787. [CrossRef]

16. Ott, I. On the Medicinal Chemistry of Gold Complexes as Anticancer Drugs. *Coord. Chem. Rev.* **2009**, *253*, 1670–1681. [CrossRef]

17. Zou, T.; Lum, C.T.; Lok, C.-N.; Zhang, J.-J.; Che, C.-M. Chemical Biology of Anticancer Gold(III) and Gold(I) Complexes. *Chem. Soc. Rev.* **2015**, *44*, 8786–8801. [CrossRef]

18. Mora, M.; Gimeno, M.C.; Visbal, R. Recent Advances in Gold–NHC Complexes with Biological Properties. *Chem. Soc. Rev.* **2019**, *48*, 447–462. [CrossRef]

19. Herrera, R.P.; Gimeno, M.C. Main Avenues in Gold Coordination Chemistry. *Chem. Rev.* **2021**, *121*, 8311–8363. [CrossRef]

20. Sadler, P.J.; Sue, R.E. The Chemistry of Gold Drugs. *Metal-Based Drugs* **1994**, *1*, 107–144. [CrossRef]

21. Azharuddin, M.; Zhu, G.H.; Das, D.; Ozgur, E.; Uzun, L.; Turner, A.P.F.; Patra, H.K. A Repertoire of Biomedical Applications of Noble Metal Nanoparticles. *Chem. Commun.* **2019**, *55*, 6964–6996. [CrossRef] [PubMed]

22. Pettenuzzo, N.; Brustolin, L.; Coltri, E.; Gambalunga, A.; Chiara, F.; Trevisan, A.; Biondi, B.; Nardon, C.; Fregona, D. Cu^{II} and Au^{III} Complexes with Glycoconjugated Dithiocarbamato Ligands for Potential Applications in Targeted Chemotherapy. *ChemMedChem* **2019**, *14*, 1162–1172. [CrossRef] [PubMed]

23. Liu, W.; Gust, R. Metal N-Heterocyclic Carbene Complexes as Potential Antitumor Metallodrugs. *Chem. Soc. Rev.* **2013**, *42*, 755–773. [CrossRef] [PubMed]

24. Stratton, M.; Ramachandran, A.; Camacho, E.J.M.; Patil, S.; Waris, G.; Grice, K.A. Anti-Fibrotic Activity of Gold and Platinum Complexes—Au(I) Compounds as a New Class of Anti-Fibrotic Agents. *J. Inorg. Biochem.* **2020**, *206*, 111023. [CrossRef]

25. Azria, D.; Blanquer, S.; Verdier, J.-M.; Belamie, E. Nanoparticles as Contrast Agents for Brain Nuclear Magnetic Resonance Imaging in Alzheimer's Disease Diagnosis. *J. Mater. Chem. B* **2017**, *5*, 7216–7237. [CrossRef]

26. Bondžić, A.M.; Janjić, G.V.; Dramićanin, M.D.; Messori, L.; Massai, L.; Parac Vogt, T.N.; Vasić, V.M. Na/K-ATPase as a Target for Anticancer Metal Based Drugs: Insights into Molecular Interactions with Selected Gold(III) Complexes. *Metallomics* **2017**, *9*, 292–300. [CrossRef]

27. Fonseca, C.; Fraqueza, G.; Carabineiro, S.A.C.; Aureliano, M. The Ca^{2+}-ATPase Inhibition Potential of Gold(I, III) Compounds. *Inorganics* **2020**, *8*, 49. [CrossRef]

28. Aikman, B.; Wenzel, M.; Mósca, A.; de Almeida, A.; Klooster, W.; Coles, S.; Soveral, G.; Casini, A. Gold(III) Pyridine-Benzimidazole Complexes as Aquaglyceroporin Inhibitors and Antiproliferative Agents. *Inorganics* **2018**, *6*, 123. [CrossRef]

29. Berrocal, M.; Cordoba-Granados, J.J.; Carabineiro, S.A.C.; Gutierrez-Merino, C.; Aureliano, M.; Mata, A.M. Gold Compounds Inhibit the Ca^{2+}-ATPase Activity of Brain PMCA and Human Neuroblastoma SH-SY5Y Cells and Decrease Cell Viability. *Metals* **2021**, *11*, 1934. [CrossRef]

30. Forestier, J. Rheumatoid arthritis and its treatment by gold salts: The results of six years' experience. *J. Lab. Clin. Med.* **1935**, *20*, 827–840.

31. Abdalbari, F.H.; Telleria, C.M. The Gold Complex Auranofin: New Perspectives for Cancer Therapy. *Discov. Oncol.* **2021**, *12*, 42. [CrossRef] [PubMed]

32. Gamberi, T.; Chiappetta, G.; Fiaschi, T.; Modesti, A.; Sorbi, F.; Magherini, F. Upgrade of an Old Drug: Auranofin in Innovative Cancer Therapies to Overcome Drug Resistance and to Increase Drug Effectiveness. *Med. Res. Rev.* **2021**, *42*, 1111–1146. [CrossRef] [PubMed]

33. Vollenbroek, F.A.; Van den Berg, J.P.; Van der Velden, J.W.A.; Bour, J.J. Phosphorus-31 [Proton] Nuclear Magnetic Resonance Investigation of Gold Cluster Compounds. *Inorg. Chem.* **1980**, *19*, 2685–2688. [CrossRef]

34. Assefa, Z.; McBurnett, B.G.; Staples, R.J.; Fackler, J.P.; Assmann, B.; Angermaier, K.; Schmidbaur, H. Syntheses, Structures, and Spectroscopic Properties of Gold(I) Complexes of 1,3,5-Triaza-7-Phosphaadamantane (TPA). Correlation of the Supramolecular Au.Cntdot. .Cntdot. .Cntdot.Au Interaction and Photoluminescence for the Species (TPA)AuCl and [(TPA-HCl)AuCl]. *Inorg. Chem.* **1995**, *34*, 75–83.

35. Cagno, V.; Gasbarri, M.; Medaglia, C.; Gomes, D.; Clement, S.; Stellacci, F.; Tapparel, C. Sulfonated Nanomaterials with Broad-Spectrum Antiviral Activity Extending beyond Heparan Sulfate-Dependent Viruses. *Antimicrob. Agents Chemother.* **2020**, *64*, e02001-20. [CrossRef]

36. Ershov, A.Y.; Martynenkov, A.A.; Lagoda, I.V.; Kopanitsa, M.A.; Zarubaev, V.V.; Slita, A.V.; Buchkov, E.V.; Panarin, E.F.; Yakimansky, A.V. Gold Glyconanoparticles Based on Aldoses 6-Mercaptohexanoyl Hydrazones and Their Anti-Influenza Activity. *Russ. J. Gen. Chem.* **2021**, *91*, 1735–1739. [CrossRef]

37. Ershov, A.Y.; Martynenkov, A.A.; Lagoda, I.V.; Yakimansky, A.V. Synthesis of 6-Mercaptohexanoylhydrazones of Mono- and Disaccharides as a Potential Glycoligands of Noble Metal Glyconanoparticles. *Russ. J. Gen. Chem.* **2020**, *90*, 1863–1868. [CrossRef]

38. Rothan, H.A.; Stone, S.; Natekar, J.; Kumari, P.; Arora, K.; Kumar, M. The FDA-Approved Gold Drug Auranofin Inhibits Novel Coronavirus (SARS-COV-2) Replication and Attenuates Inflammation in Human Cells. *Virology* **2020**, *547*, 7–11. [CrossRef]

39. Zacheo, A.; Hodek, J.; Witt, D.; Mangiatordi, G.F.; Ong, Q.K.; Kocabiyik, O.; Depalo, N.; Fanizza, E.; Laquintana, V.; Denora, N.; et al. Multi-Sulfonated Ligands on Gold Nanoparticles as Virucidal Antiviral for Dengue Virus. *Sci. Rep.* **2020**, *10*, 9052. [CrossRef]
40. Gil-Moles, M.; Türck, S.; Basu, U.; Pettenuzzo, A.; Bhattacharya, S.; Rajan, A.; Ma, X.; Büssing, R.; Wölker, J.; Burmeister, H.; et al. Metallodrug Profiling against SARS-CoV-2 Target Proteins Identifies Highly Potent Inhibitors of the S/ACE2 Interaction and the Papain-like Protease PL^pro. *Chem. A Eur. J.* **2021**, *27*, 17928–17940. [CrossRef]
41. Ouji, M.; Bourgeade-Delmas, S.; Álvarez, F.; Augereau, J.; Valentin, A.; Hemmert, C.; Gornitzka, H.; Benoit-Vical, F. Design, Synthesis and Efficacy of Hybrid Triclosan-Gold Based Molecules on Artemisinin-Resistant Plasmodium Falciparum and Leishmania Infantum Parasites. *ChemistrySelect* **2020**, *5*, 619–625. [CrossRef]
42. Ouji, M.; Barnoin, G.; Fernández Álvarez, Á.; Augereau, J.-M.; Hemmert, C.; Benoit-Vical, F.; Gornitzka, H. Hybrid Gold(I) NHC-Artemether Complexes to Target Falciparum Malaria Parasites. *Molecules* **2020**, *25*, 2817. [CrossRef] [PubMed]
43. Feng, L.; Pomel, S.; Latre de Late, P.; Taravaud, A.; Loiseau, P.M.; Maes, L.; Cho-Ngwa, F.; Bulman, C.A.; Fischer, C.; Sakanari, J.A.; et al. Repurposing Auranofin and Evaluation of a New Gold(I) Compound for the Search of Treatment of Human and Cattle Parasitic Diseases: From Protozoa to Helminth Infections. *Molecules* **2020**, *25*, 5075. [CrossRef]
44. Koko, W.S.; Jentzsch, J.; Kalie, H.; Schobert, R.; Ersfeld, K.; Al Nasr, I.S.; Khan, T.A.; Biersack, B. Evaluation of the Antiparasitic Activities of Imidazol-2-ylidene–Gold(I) Complexes. *Arch. Pharm.* **2020**, *353*, e1900363. [CrossRef] [PubMed]
45. Miyamoto, Y.; Aggarwal, S.; Celaje, J.J.A.; Ihara, S.; Ang, J.; Eremin, D.B.; Land, K.M.; Wrischnik, L.A.; Zhang, L.; Fokin, V.V.; et al. Gold(I) Phosphine Derivatives with Improved Selectivity as Topically Active Drug Leads to Overcome 5-Nitroheterocyclic Drug Resistance in *Trichomonas vaginalis*. *J. Med. Chem.* **2021**, *64*, 6608–6620. [CrossRef]
46. Tu, X.; Tan, X.; Qi, X.; Huang, A.; Ling, F.; Wang, G. Proteome Interrogation Using Gold Nanoprobes to Identify Targets of Arctigenin in Fish Parasites. *J. Nanobiotechnol.* **2020**, *18*, 32. [CrossRef]
47. Minori, K.; Rosa, L.B.; Bonsignore, R.; Casini, A.; Miguel, D.C. Comparing the Antileishmanial Activity of Gold(I) and Gold(III) Compounds in *L. amazonensis* and *L. braziliensis* In Vitro. *ChemMedChem* **2020**, *15*, 2146–2150. [CrossRef]
48. Cagno, V.; Andreozzi, M. Broad-Spectrum Non-Toxic Antiviral Nanoparticles with a Virucidal Inhibition Mechanism. *Nat. Mater.* **2018**, *17*, 195–203. [CrossRef]
49. Cagno, V.; Tseligka, E.D.; Jones, S.T.; Tapparel, C. Heparan Sulfate Proteoglycans and Viral Attachment: True Receptors or Adaptation Bias? *Viruses* **2019**, *11*, 596. [CrossRef]
50. Jones, S.T.; Cagno, V.; Janeček, M.; Ortiz, D.; Gasilova, N.; Piret, J.; Gasbarri, M.; Constant, D.A.; Han, Y.; Vuković, L.; et al. Modified Cyclodextrins as Broad-Spectrum Antivirals. *Sci. Adv.* **2020**, *6*, eaax9318. [CrossRef]
51. Guven, Z.P.; Silva, P.H.J.; Luo, Z.; Cendrowska, U.B.; Gasbarri, M.; Jones, S.T.; Stellacci, F. Synthesis and Characterization of Amphiphilic Gold Nanoparticles. *J. Vis. Exp.* **2019**, *149*, 58872. [CrossRef] [PubMed]
52. Paloque, L.; Hemmert, C.; Valentin, A.; Gornitzka, H. Synthesis, Characterization, and Antileishmanial Activities of Gold(I) Complexes Involving Quinoline Functionalized N-Heterocyclic Carbenes. *Eur. J. Med. Chem.* **2015**, *94*, 22–29. [CrossRef] [PubMed]
53. Zhang, C.; Bourgeade Delmas, S.; Fernández Álvarez, Á.; Valentin, A.; Hemmert, C.; Gornitzka, H. Synthesis, Characterization, and Antileishmanial Activity of Neutral N-Heterocyclic Carbenes Gold(I) Complexes. *Eur. J. Med. Chem.* **2018**, *143*, 1635–1643. [CrossRef] [PubMed]
54. Straimer, J.; Gnädig, N.F.; Stokes, B.H.; Ehrenberger, M.; Crane, A.A.; Fidock, D.A. *Plasmodium Falciparum* K13 Mutations Differentially Impact Ozonide Susceptibility and Parasite Fitness In Vitro. *mBio* **2017**, *8*, e00172-17. [CrossRef] [PubMed]
55. Jortzik, E.; Farhadi, M.; Ahmadi, R.; Tóth, K.; Lohr, J.; Helmke, B.M.; Kehr, S.; Unterberg, A.; Ott, I.; Gust, R.; et al. Antiglioma Activity of GoPI-Sugar, a Novel Gold(I)–Phosphole Inhibitor: Chemical Synthesis, Mechanistic Studies, and Effectiveness In Vivo. *Biochim. Biophys. Acta (BBA)-Proteins Proteom.* **2014**, *1844*, 1415–1426. [CrossRef] [PubMed]
56. Ielo, I.; Rando, G.; Giacobello, F.; Sfameni, S.; Castellano, A.; Galletta, M.; Drommi, D.; Rosace, G.; Plutino, M.R. Synthesis, Chemical–Physical Characterization, and Biomedical Applications of Functional Gold Nanoparticles: A Review. *Molecules* **2021**, *26*, 5823. [CrossRef] [PubMed]

Review

Native Protein Template Assisted Synthesis of Non-Native Metal-Sulfur Clusters

Biplab K. Maiti [1],* and José J. G. Moura [2]

[1] Department of Chemistry, School of Sciences, Cluster University of Jammu, Canal Road, Jammu 180001, India

[2] LAQV, REQUIMTE, Department of Chemistry, NOVA School of Sciences and Technology (FCT NOVA), 2829-516 Caparica, Portugal; jose.moura@fct.unl.pt

* Correspondence: biplabmaiti@clujammu.ac.in

Abstract: Metalloenzymes are the most proficient nature catalysts that are responsible for diverse biochemical transformations introducing excellent selectivity and performing at high rates, using intricate mutual relationships between metal ions and proteins. Inspired by nature, chemists started using naturally occurring proteins as templates to harbor non-native metal catalysts for the sustainable synthesis of molecules for pharmaceutical, biotechnological and industrial purposes. Therefore, metalloenzymes are the relevant targets for the design of artificial biocatalysts. The search and development of new scaffolds capable of hosting metals with high levels of selectivity could significantly expand the scope of bio-catalysis. To meet this challenge, herein, three native scaffolds: [1Fe-4Cys] (rubredoxin), [3Fe-4S] (ferredoxin), and [$S_2MoS_2CuS_2MoS_2$]-ORP (orange protein) protein scaffolds are case studies describing templates for the synthesis of non-native monomeric to mixed metal–sulfur clusters, which mimic native Ni containing metalloenzymes including [Ni-Fe] Hydrogenase and [Ni-Fe] CO Dehydrogenase. The non-native metal-substituted metalloproteins are not only useful for catalysis but also as spectroscopic probes.

Keywords: designed metalloproteins; models of [Ni-Fe]-hydrogenase and [Ni-Fe]-CODH; orange-protein and spectroscopic probes

Citation: Maiti, B.K.; Moura, J.J.G. Native Protein Template Assisted Synthesis of Non-Native Metal-Sulfur Clusters. *BioChem* **2022**, *2*, 182–197. https://doi.org/10.3390/biochem2030013

Academic Editors: Manuel Aureliano, M. Leonor Cancela, Célia M. Antunes and Ana Cristina Rodrigues Costa

Received: 21 March 2022
Accepted: 30 June 2022
Published: 1 August 2022

Publisher's Note: MDPI stays neutral with regard to jurisdictional claims in published maps and institutional affiliations.

1. Introduction

Nature has evolved in order to utilize metal ions and/or metal clusters within protein scaffolds to build up metalloproteins that accomplish diverse chemical reactions enabling to sustain of life [1–4]. The versatility of the metals and biological ligands available in proteins is amazing. The same metal (with a set of conserved amino acids as ligands) may show different electronic/physical properties, performing a wide range of biological roles in different metalloproteins [5–7]. Nature utilizes a range of different metals and recruits the correct metal into proper protein environments to execute selective functions [5–7]. The nuclearity of metal-cofactors varies from monomeric to multimeric. Monomeric metalloenzymes are well studied, such as cupredoxin [8], rubredoxin [9], cytochrome P450 [10,11], and molybdenum-enzymes [12], which are involved in a variety of biochemical transformations, and with relevant roles in electron transfer processes. Furthermore, many biochemical transformations occurred by a variety of complex metalloenzymes such as nitrogen-fixing nitrogenases [13,14], photosystem [15,16], hydrogenases [17,18], and carbon monoxide-dehydrogenase (CODH) [19,20].

However, many enzymes show intrinsic promiscuity [21,22] for various forms of chemical reactions, whereas other activities are obtained by only a small alteration of their active site or protein environment [23]. The diversity of promiscuous enzymatic activity can be expanded by the incorporation of a variety of metallic ions at the active sites of metalloproteins, catalyzing a wide range of chemical transformations [2,24–26]. Handling of the metal-binding site of metalloprotein is usually aimed for two main reasons: (i) to replicate the active site of other native metalloenzymes, and (ii) to design spectroscopic probes for elucidating the structure and function of native metalloenzymes. Herein, three native protein scaffolds, [1Fe-4Cys] (rubredoxin) [9], [3Fe-4S] (ferredoxin) [27,28] and [$S_2MoS_2CuS_2MoS_2$]-ORP (orange protein) [29,30] polypeptides are considered as templates and have proven a handy platform for the synthesis of non-native metal–sulfur or mixed-metal–sulfur clusters with novel functions. The protein-assisted metal–sulfur cluster synthesis was started in the 1980s to investigate the structures and functions of metalloproteins [31–33] and then explored in many metalloproteins [34–36]. Indeed, Rubredoxin (Rd) is the simplest and smallest iron–sulfur protein possessing one Fe that is replaced by a series of non-native metal ions to study the biochemical properties of Rd [9] and to design the model compounds, such as Ni-Rd, that are considered as a model of [Ni-Fe]-Hydrogenase [37,38].

In the case of [3Fe-4S]-ferredoxin, the vacant Fe-site is filled up by a variety of metal ions to yield a wide range of hetero-metal–sulfur cubane clusters ([M,3Fe-4S]). Interestingly, designed [Ni,3Fe-4S]-ferredoxin mimics the native [Ni-Fe] carbon monoxide dehydrogenase [39]. The third protein, ORP, Ref. [29] is used as a template for the synthesis of hetero-metal-clusters as spectroscopic probes by replacing the diamagnetic Cu^I from [$S_2MoS_2CuS_2MoS_2$]$^{3-}$ of ORP, enlarging the scope of the initial studies [40]. Therefore, this review focuses on non-native metal ions that are incorporated in active sites by chemical manipulation of protein template, aiming at the synthesis of derivatives that may be described as either model enzymes or spectroscopic probes.

2. Covalently vs. Non-Covalently Coordinated Metal-Cofactors

In metalloenzymes, metal cofactors are bound to certain amino acid residues, which are necessary to drive the many biochemical transformations. Naturally occurring metalloproteins possess a native metallocofactor that can be attached to protein templates through covalent or non-covalent interactions by a group of side chain amino acid residues. The covalently attached metallocofactors are commonly found in many metalloproteins such as rubredoxin [9], ferredoxin [27], molybdoenzymes [12], nitrogenase [13], and hydrogenase [17]. In contrast, non-covalently attached metallocofactors are observed in a very limited number of metalloproteins, such as orange protein (ORP) [29] and Mo-Fe protein [41]. However, covalent anchoring metallocofactors are more strongly embedded in protein scaffolds than non-covalent anchoring metallocofactors, but both types of metal-cofactors are directly tuned by protein scaffolds. Furthermore, unlike covalent assembly, non-covalent (supramolecular) assembly can allow the entry and exit of the guest molecule in the host cavity reversibly. The designed information of molecular structure is stored in the host-cavity (protein template) that builds the guest structure through non-covalent interactions, including hydrogen-bonding, ion-pair, and hydrophobic interactions [42,43]. Inspired by Nature, the non-native metal ions have been recruited into protein scaffolds by non-covalent or covalent chemical ways to design artificial metalloenzymes.

3. Iron–Sulfur Proteins

Iron–sulfur ([Fe-S]) clusters are the utmost ancient and ubiquitous biological inorganic cofactors that are involved in a wide range of biochemical processes, including electron transfer, gene regulation, and catalysis [44–46]. [Fe-S] clusters show a variety of frameworks in biology and the most common structural types are [1Fe-4Cys] in rubredoxin [9], [2Fe-2S] in plant ferredoxin [4,47], and [4Fe-4S] in bacterial ferredoxin [48]. In addition, [3Fe-4S] type clusters are found in biology, structurally related to [4Fe-4S] cores, namely in ferredoxins and several complex metalloproteins [27,28]. Apart from the classic [Fe-S]-clusters, the unique, bigger, and highly complex [Fe-S]-clusters are found in many proteins such as shiroheme-[4Fe-4S]-cluster in sulfite reductase [49–51], [Mo-7Fe-9S] (FeMo-cofactor) and [8Fe-7S] (P-cluster) clusters in nitrogenase [52], a unique H-cluster in [Fe-Fe] hydrogenases [53], a hetero-metal cluster in [Ni-Fe] hydrogenase [17] and [Ni,4Fe-5S] cluster in carbon monoxide dehydrogenase [54]. Furthermore, a complex [8Fe-9S] cluster in the ATP-dependent reductase from *Carboxydothermus hydrogenoformans* [55] and a non-cubane [4Fe-4S] cluster in the heterodisulfide reductase from methanogenic archaea are observed [56]. Due to structural/functional variability, iron–sulfur clusters are attractive templates to be used for the design of artificial metalloproteins/enzymes, aiming at the elucidation of the structure and function of the intricate systems and possibly other performances such as catalysts. Therefore, in this review, we are focusing on two types of iron–sulfur proteins, mononuclear-rubredoxin, and trinuclear-[3Fe-4S]-ferredoxin.

3.1. Overview of Rubredoxin

Rubredoxins (Rds), a small (5.6 kDa) and simplest case among iron–sulfur proteins, are observed mainly in anaerobic bacteria and archaea [9,57] and one group of eukaryotes (photosynthetic algae and plants) [58,59]. The active site of Rd possesses one iron atom, tetrahedrally ligated by four cysteine residues from two $-CX_2C-X_n-CX_2C-$segments in a polypeptide chain (Figure 1). A variation of this binding site contains two adjacent cysteine-binding modes ($-CX_2C-X_n-CC-$) in a protein, namely desulforedoxin [60]. Based on metal composition, Rds are mainly two types which are single Fe-centre systems (rubredoxins and flavo-rubredoxins) and complex Fe-centre systems, wherein the Rd center is coupled with other types of iron centers (rubrerythrins, nigerythrins, desulfoferrodoxins, and desulforedoxins) [9,61]. The biological functions of Rds are still unclear, but it presumes to participate in the e^- transfer process cycling between ferrous and ferric forms, in many biochemical processes, which are fatty acid metabolism, detoxification of reactive oxygen species [62,63], and carbon fixation [64]. The redox potentials of all types of Rds fall in the range from -0.1 V to $+0.1$ V (vs. NHE) in spite of the same prosthetic group, FeS_4 [9,61]. It is well established that the redox potential of Rds is highly influenced by many parameters, which are Fe-S bonds, hydrogen bondings [65,66], variation in solvation, electrostatic interaction, and dipole moment. The crystal structures of reduced and oxidized Rds reveal that the average Fe-S bond distance in the oxidized Rd is smaller than the reduced Rd ($d_{Fe^{III}-S}$ = 2.25–2.30 Å vs. $d_{Fe^{II}-S}$ = 2.30–2.40 Å) (Figure 1) [67].

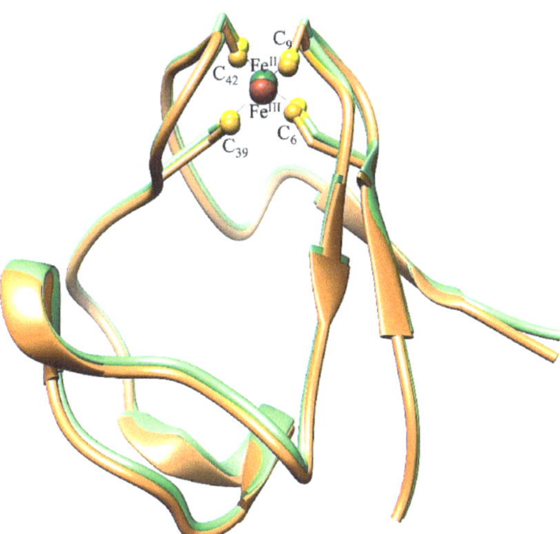

Figure 1. Superimposed of X-ray structures of native Rds from *Clostridium pasteurianum*; oxidized Fe^{III}-Rd (light brown ribbon with red Fe^{III} ball) (PDB: 1FHH) and reduced Fe^{II}-Rd states (light green ribbon with green Fe^{II} ball) (PDB: 1FHM).

Furthermore, the protein fold of Rd is remarkably stable under aerobic conditions, in a wide range of pH, solvents, and temperatures, and also stable toward mutagenesis [68]. The native iron in Rd can be replaced by a wide range of metal ions, including $^{57}Fe^{II}$, Co^{II}, Ni^{II}, Cu^{II}, Zn^{II}, Cd^{II}, Hg^{II}, Ga^{III}, In^{III}, and Mo^{VI} [9,38,69–72]. A series of non-native metal substituted Rd variants have been studied to explain the structural, electronic, and magnetic properties of metal ions in active sites and are also useful as model systems. Indeed, The Cu-Rd derivative enlarges our knowledge of the redox chemistry interplay between Cu^I and Cu^{II} in a sulfur-rich protein environment [70,71]. Interestingly, cysteine-rich Cu^{II}-Rd is a remarkably stable compound under dark and argon, indicating thermodynamically unfavorable to "Cu-thiol-auto-reduction" [71]. In addition, molybdenum was also incorporated in apo-rubredoxin to afford the insertion of molybdenum in a cysteinyl coordination sphere, which is of interest as a model complex for resting or active state of molybdoenzymes, which hold a remarkable place in bioinorganic chemistry, because they perform a series of metabolic reactions in the carbon, nitrogen, and sulfur biocycles. However, the designed MoO_2-Rd derivative promotes the oxygen atom transfer reactions, such as the oxidation of arsenite to arsenate [72]. Among these, designed Ni-Rd is mostly studied as a model of [Ni-Fe]-hydrogenase [37,38,73].

3.1.1. Ni-Substituted Rd: Model of [Ni-Fe]-Hydrogenase

Nature designed an efficient enzyme, Hydrogenases, that reversibly cleave hydrogen, a clean and alternative chemical feedstock to non-renewable energy sources. Hydrogenases have three distinct families, which are [Fe-Fe], [Ni-Fe], and Fe (only) hydrogenases [17,74]. Among them, the bimetallic active site in [Ni-Fe]-hydrogenases are covalently ligated by four cysteine residues. Two cysteines bridge the two metal centers (Ni and Fe), and the other two cysteines are terminally coordinated to the Ni-atom. In addition, unusual biological ligands, CO and CN^- are bonded to Fe-atom in the active site and make them fascinating examples of 'organometallic' cofactors [75–77] (Figure 2A). The Ni-atom is believed to be the center of activity for hydrogen evolution. The extensive study of Hydrogenases has a central focus on structure/function aspects for a future protein-based H_2-evolving technology. Parallelly, the bio-inspired small synthetic model systems have extensively progressed for hydrogen conversion catalysts but are not yet comparable to the naturally occurring enzymes [78–81]. It is known that the protein matrix plays a vital role in tuning the activity of the inorganic active site, but direct enzymological studies may be hampered by biological complications/complexities due to its large size with complex catalytic cofactor and toxicity toward the dioxygen molecule. Therefore, size scale intermediates between native enzymes and organometallic model compounds can provide many advantages for both regimes [82–84]. In this perspective, small stable proteins such as rubredoxin and its metal substituted derivatives are available for modeling the active site of bacterial [Ni-Fe]-Hydrogenases.

Ni-substituted Rd (Ni-Rd), an air-stable derivative (Figure 2B), is one such model showing identical structural features at the primary coordination sphere of Ni-fragment of native [Ni-Fe]-hydrogenase (Figure 2A) and native-enzyme-like activity including the evolution of hydrogen, deuterium-proton exchange, and inhibition of hydrogen-evolving activity by carbon monoxide [38]. The characterization of hydrogen-evolving Ni-Rd activity has been significantly expanded during the last decade [85–87]. The first study by Moura and Co-workers reported that recombinant Ni-Rd from *Desulfovibrio* exhibited lower hydrogen-evolving activity, but recently, Shafaat and Co-workers have reported that recombinant Ni-Rd from *Desulfovibrio desulfuricans* ATCC 27774 displays light initiated H_2 production with a high turnover frequency of about 20–100 s^{-1} at 4 °C in solution upon electrochemical study. The electrocatalytic over-potential is about 550 mV, which is comparable to native [Ni-Fe]-hydrogenase at the same condition [85,87]. Furthermore, the covalent attachment of Ni-Rd to the graphite electrode, coupling through amide bond formation shows stable H_2-evolving activity for many weeks with a higher turnover number of about 6700, but it shows lower turnover frequency due to the slower interfacial electron transfer (ET) rates that modulates the catalytic rate [88]. Ni-Rd is an ideal candidate for studying the molecular mechanism of the native [Ni-Fe]-hydrogenase. A combined experimental and theoretical data indicate that the proton-coupled electron transfer is a vital step for catalysis where Cys-thiolates act as the site for protonation, suggesting a similar mechanism as native [37,86]. The secondary coordination sphere of metalloprotein plays a significant role in the catalytic cycle. In order to understand the effect of the protein environment on the H_2-activity of Ni-Rd, a group of mutated Rds is generated, which influences the H_2-evolving activity, suggesting the modulation of the H-bonding network at the vicinity of the active site. The result indicates that Cys35 is the primary site for protonation during catalysis [73]. Moreover, Ni-Rd is generated in vivo, and it is indistinguishable from chemically substituted Ni-Rd, presenting a structural and functional replica of native enzymes [89].

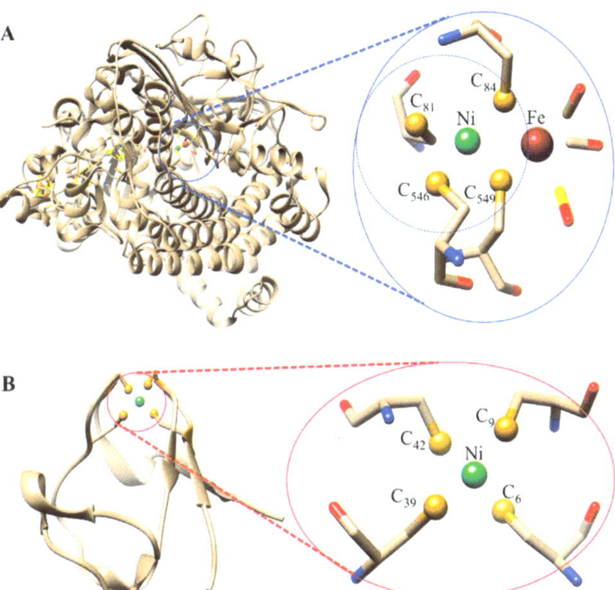

Figure 2. The Ni-Rd model shows identical structural features at the primary coordination sphere of Ni-fragment of native [Ni-Fe]-hydrogenase. (**A**) Crystal structures of [Ni-Fe]-Hydrogenase from *Desulfovibrio vulgaris Miyazaki F* (PDB:1H2R), highlighted the [Ni-Fe] active site with cysteine coordination and (**B**) Crystal structure of Ni-Rd from *Desulfovibrio gigas* (PDB:1R0J), highlighted the Ni-site with cysteine coordination.

Ni-Rd is also extended to design hybrid catalysts for light-driven H_2 evolution, namely "solar fuel", which is an emerging research area [90–93]. In order to meet this, a ruthenium-based chromophore is covalently connected to mutated Ni-Rd through free Cys31 residue to build a hybrid enzyme, Ru,Ni-Rd, that is capable of photo-induced hydrogen production [87,94]. The hydrogen generation rate highly depends on the distance between Ru and Ni centers in Ru,Ni-Rd suggesting the intramolecular electron transfer in catalysis. A series of Rd variants are designed as probes where free cysteine residue is located at different positions 17, 31, 38, and 45 to understand the impact of chromophore attachment site on hydrogen evolving activity of Ru,Ni-Rd [87]. The hydrogen production efficiency is lowest when the Ru-chromophore is the longest distance from the Ni-active site. This approach is ideal for the manufacture of solar fuels [87]. The modification of metal-site in rubredoxin will be considered great progress and a breakthrough in the biosynthetic inorganic chemistry field. Therefore, the outstanding results and progress in this research would certainly have a great impact on the future as a potential alternative fuel source.

3.1.2. Spectroscopic Probes-M-Rd

A number of metal-substituted Rd derivatives are available and are also useful for the study of various spectroscopies due to their rich optical and magnetic nature in order to understand the structure and function of Rds [9]. Indeed, the non-native ^{57}Fe-Rd is employed as a Mössbauer probe for the study of the oxidation state of the Fe center as well as the covalent nature of the Fe-S bond. Mössbauer spectroscopy study indicates larger Fe-S covalency nature in Desulforedoxin over Rd [95,96]. The other spectroscopic probe is NMR, which gives the information about the cysteine coordination at Fe-site in Rd but paramagnetic Fe causes broadening of ^{1}H-NMR [97]. In order to overcome this paramagnetic NMR, the native Fe in Rd is replaced by a diamagnetic Zn^{II} [98,99]. The NMR study of Zn-Rd shows well-defined as well as well-resolved NMR peaks that indicate a similar structure to the native structure [100,101]. Likewise, Cd substituted Rd is more advantageous for ^{113}Cd-NMR study to gain more structural information [102,103]. The Cd-Rd is extensively used as a ^{113}Cd-NMR probe for studying the metal–cysteine ligation in various metalloproteins [104]. The potential redox value of Rds is highly affected by the six conserved H-bonding interactions which are observed between cysteine–sulfur and vicinal amide-protons [85]. The NMR study of ^{113}Cd-substituted Rd-Cp and WT-Rd-Cp shows a large ^{1}H chemical shift of amide proton in the vicinity of the active site (NH—S), suggesting the variation of the redox potential [105]. The small M-Rd derivative is an efficient NMR probe to study the Fe-S—HN interaction in other [M,Fe-S] proteins.

Interestingly, the native Fe-Rd is also chosen as a paramagnetic NMR probe to be useful for the protein–protein interaction study [98]. Indeed, two proteins, Fe-Rd and cytochrome c3 (cyt c3), are considered binding partners. The paramagnetic probe Fe-Rd maps the interface of the target protein, cyt c3, by NMR study. The NMR study identifies the heme methyl groups in cyt c3 that participate in the binding surface interface of the Rd-cyt c3 complex [98]. This is a valuable approach to extending the study to other protein partners.

3.2. Overview of [3Fe-4S] Ferredoxin

Widespread distribution and multiple roles of iron–sulfur clusters with variable structure and oxidation states have been increasingly disclosed. [Fe-S]-proteins show diverse functions, which are electron transfer, nitrogen fixation, photosynthesis, enzymatic catalysis, signaling, respiration, gene regulation, and DNA repair and replication [106–110]. Ferredoxins (Fds) are an important class of [Fe-S] proteins and have different sub-classes according to their composition of iron–sulfur, including [2Fe-2S], which is known as plant–Fd [111,112], and [3Fe–4S] and [4Fe–4S] centers are known as bacteria ferredoxin [27,28]. Like Rd, the redox chemistry of [3Fe-4S] is also highly influenced by pH because it has an inherent tendency for protonation [7,113]. This protonation chemistry has been found in all [3Fe-4S]-ferredoxin such as [3Fe–4S] ferredoxins from *P. furiosus* [114], *D. gigas* (Fd II) [115], and beef heart aconitase [116], and [7Fe-8S] ([3Fe– 4S] and [4Fe–4S]) ferredoxins from *A. Vinelandii* (Fd I) [117], and *D. africanus* (Fd III) [118]. The structure of the cuboidal [3Fe–4S] cluster is an open-faced crown-like structure (Figure 3) with three inorganic μ_2-S atoms, which are active centers for protonation [113]. This chemistry is flourished for the synthesis of hetero-metal cluster synthesis by introducing the non-native metal into the open-faced site (below).

A common [4Fe–4S] cluster is covalently attached to a protein scaffold with a typical cysteine binding motif: $-Cys-X_2-Cys-X_2-Cys-X_n-Cys-$ or Cys-X-X-Cys-X-X-Cys. The other type, the [3Fe-4S] cluster, differs in only one Fe-center missing at the corner site of the cubane [4Fe-4S] cluster [119] and is generally bound in a polypeptide chain with three cysteine residues. Interestingly, both trimeric cuboidal and teturameric cubane clusters are reversibly interconverted. The first protein-bound [3Fe-4S] cluster was found in aconitase in 1984 [120]. The [3Fe-4S] is an inactive form of aconitase, and it is easily interconverted to form an active [4Fe–4S] cluster [120]. Indeed, an easy interconversion between [3Fe–4S] and [4Fe–4S] clusters is observed in *Pyrococcus furiosus* ferredoxin, where one cysteine out of four is replaced by aspartate residue (Figure 3) [121]. In addition, in *Desulfovibrio africanus ferredoxin III*, one of two [4Fe–4S] clusters coordinates with three Cys residues, and it readily converts to the [3Fe–4S] cluster [122]. The fourth iron is ligated by abiological water and hydroxide ligands [123] or protein-derived carboxylate ligands [124].

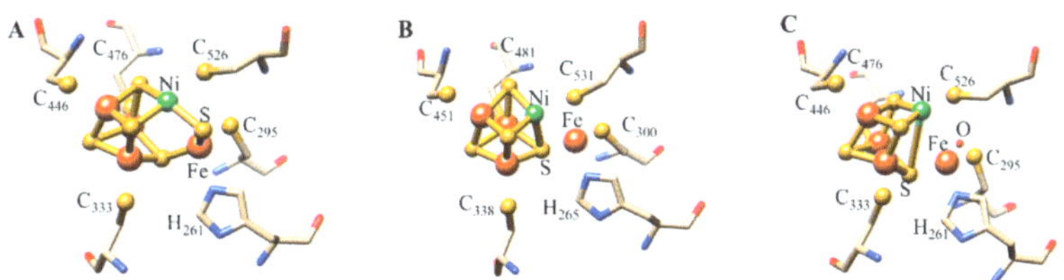

Figure 3. Inter-conversion between [3Fe-4S] and [4Fe-4S].

This conversion has suggested a possible route of recruiting other metal ions into the [3Fe-4S] core to yield a series of novel heteromeric [M,3Fe-4S] type clusters. The construction of [M,3Fe-4S] clusters is preceded by reactions with exogenous metal ions. Indeed, the first result was the formation of the [Co,3Fe-4S]$^{2+}$ cluster that was recognized in 1986 [125]. Since that time, the [3Fe-4S] core has served as a precursor of the formation of [M,3Fe-4S]. Therefore, a series of metal ions are incorporated into the [3Fe-4S] core to yield various hetero-metal–sulfur clusters [M,3Fe-4S] within the protein environment [36,126–128].

3.2.1. Ni-Incorporated in [3Fe-4S]-Fd: Model of Ni-Containing CODH

CO dehydrogenases (CODHs) catalyze the reversible and selective oxidation of CO to CO_2 [129,130]. Nature evolves two types of CODHs: (1) Mo-Cu containing CODHs found in aerobic organisms [131,132], and (2) [Ni-Fe] containing CODHs found in anaerobic organisms [133,134]. Anaerobic bacteria such as *Carboxydothermus hydrogenoformans* (CODH$_{Ch}$), *Moorella thermoacetica* (CODH$_{Mt}$), and *Rhodospirillum rubrum* (CODH$_{Rr}$) use CO as a carbon source [135]. The crystal structures of the active sites of [Ni-Fe]-CODH have been reported and harbor a [Ni,4Fe-4S–OHx], [Ni,4Fe-5S] (μ_2-sulfide covalently bridges the Ni-atom and the distal Fe-atom) and [Ni,4Fe-4S] cluster (Figure 4) [135,136].

Figure 4. Structures of [Ni-Fe]-CODH from different sources, [Ni,3Fe-5S] (**A**) from CODH$_{Ch}$ (PDB file 1SU6), [Ni,3Fe-4S] (**B**) from CODH$_{Rr}$ (PDB file 1JQK) and [Ni,3Fe-4S-O] (**C**) from CODH-II$_{Ch}$ (PDB file 3B53).

A wide range of metal substituted [3Fe-4S]-ferredoxin derivatives are reported. Among them, [Ni-3Fe-4S] ferredoxin can be considered as a simplified protein-based model of the native [Ni-Fe]-CODH [39]. Indeed, *Pyrococcus furiosus* ferredoxin (*Pf*-Fd), a small e$^-$ transfer protein contains a cubane [4Fe-4S] cluster that facile interconverts to [3Fe-4S] cluster under suitable experimental conditions due to the presence of non-cysteinyl ligation (mainly asparted) at the specific one Fe site [121,137–139]. The apo-*pf*-Fd protein has been recruited with a [Ni,3Fe-4S] cluster that mimics the active site core of the native CODH. Recently, Lewis et al. reported that the [Ni,3Fe-4S] cluster is reconstituted in *pf*-Fd to generate a protein-based simplified model for [Ni-Fe]-CODH that shows reversible e- transfer process and binding of both CO (as substrate) and CN$^-$ (as inhibitor) of CODH [39].

3.2.2. Spectroscopic Probes-[M,3Fe-4S] Ferredoxin

A wide range of non-native metals such as Cu [140], Mn [127], Ni [36,115], Co [115,125,127], Zn [118,141,142], Cd [115,118] and Ag [143] are incorporated into the vacant space in the cuboidal [3Fe-4S] core for the studies of the optical, electronic, magnetic, and redox properties of mixed metal cluster, [M,3Fe-4S]. Indeed, [Cd,3Fe-4S] cluster shows the ground state spin value of S = 2 whereas [Cu,3Fe-4S] shows S = 1/2 [128]. Furthermore, the redox potentials of [Cd,3Fe-4S]$^{2+/+}$ and [Cu,3Fe-4S]$^{2+/+}$ couples are -470 mV and +190 mV (vs. NHE) respectively [128], suggesting that the ground spin state and redox potential are functions of the incorporation of metal ions in the [3Fe-4S] fragment. The [3Fe-4S]^0 core has two sites, including a delocalized Fe^{II}•Fe^{III} pair and a localized Fe^{III} site. Upon addition of Zn^{II} ion in [3Fe-4S] fragment, the formation of [Zn,3Fe-4S]$^+$ shows unusual hyperfine interaction with spin state of 5/2 using Mossbauer spectra. In contrast, Mossbauer spectra of three Fe^{II} sites in [Ni,3Fe-4S]$^+$ show the same quadrupole splitting with the same isomer shifts suggesting that three Fe^{II} sites are delocalized rather than localized Fe^{II} sites [141]. Therefore, small [M,3Fe-4S] derivatives are efficient spectroscopic probes that can be applied to other cubane iron–sulfur proteins.

4. Overview of Orange Proteins

Orange protein (ORP), a monomeric small (~12 kDa) protein possesses a novel hetero-metallic cofactor of the type [S$_2$MoS$_2$CuS$_2$MoS$_2$]$^{3-}$ that was first isolated from sulfate-reducing bacteria, *Desulfovivrio gigas* in 2000, and still, the function is unknown [29,30]. Now, the ORP encoded gene is found in many anaerobic bacteria [47,144]. The mixed metal–sulfur cluster is non-covalently attached to the protein matrix through hydrophobic and electrostatic interactions [29]. The ORP-Cu cluster is stable in solution for a long time, but the cofactor can be released from the host cavity by disrupting the non-covalent interaction in an irreversible process due to the self-rearrangement to yield a larger cluster that cannot be fit into the protein cavity [145]. Unfortunately, the holo-ORP is crystalized as apo-ORP without a metallocofactor, and the crystal structure of apo-ORP is shown in Figure 5 [146]. Furthermore, NMR data of apo-ORP [147,148] reveal a metal cluster-binding region in D21–A27, H53–N58, and L72–F81 amino acid residues. However, the exact metal cofactor binding region of ORP remains uncharacterized yet. This unusual mixed metal–sulfide cofactor in ORP is the result of the Mo/Cu antagonism that was first found in ruminants [149] and is now being utilized in Wilson's diseases and various states of cancer for the applications of anticopper therapies [149,150].

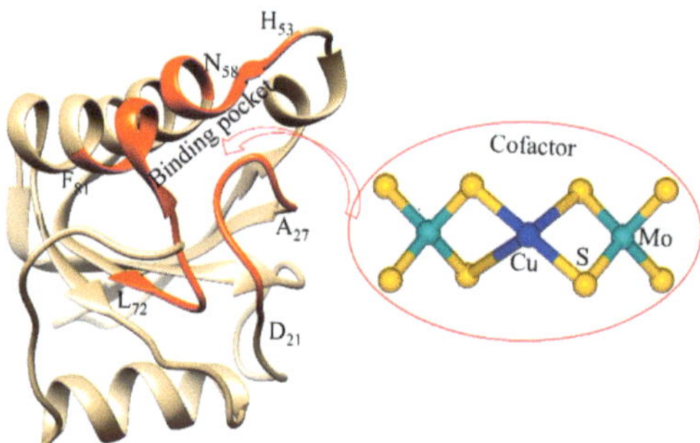

Figure 5. Crystal structure of apo-ATCUN-ORP (PDB: 2WFB) where metal-cofactor [S$_2$MoS$_2$CuS$_2$MoS$_2$]$^{3-}$ (EXAFS structure) binding region amino acid residues are highlighted with red color in ribbon.

Protein-assisted syntheses of metal-cluster methodologies are quite inspired by the large experience of the Mour's group acquired in the past on self-assembly of iron–sulfur centers and hetero-metal sites in iron–sulfur proteins (described above) [125,142]. During the reconstitution procedure, firstly CuCl$_2$ was added, followed by TTM (Tetrathiomolybdate; $MoS_4{}^{2-}$), and the $[Cu(TTM)_2]^{3-}$-ORP was obtained with the exact composition of Mo:Cu stoichiometric ratio of 2:1, but in reverse additions, different mixed metal–sulfur composition were obtained. This result can be concluded since TTM has a tendency to absorb protein by either non-covalently interaction or charge interaction with positive charge amino acid; consequently, TTM does not go smoothly towards the protein cavity site. Therefore, the addition of TTM, followed by the addition of $CuCl_2$, makes TTM sulfur ligands bind more copper ions resulting in the formation of different types of non-native Mo/Cu composition that is not allowed into the apo-ORP cavity. We can also propose that, in the first step of metal-cofactor formation, copper goes into the cavity region of apo-ORP and promotes the entry of TTM due to the nature of Mo-Cu antagonism, suggesting a driving force for this process. In an alternative way, we can propose that the apo-ORP protein does not promote a specific cavity, but upon the addition of $CuCl_2$ and TTM into apo-ORP, the protein makes a cavity for the accommodation of the discrete a native mixed metal–sulfur cluster. The simple inorganic cofactor, $[S_2MoS_2CuS_2MoS_2]^{3-}$ [151], in the solution is no longer stable and itself rearranges to afford a bigger cluster, but in the presence of a protein matrix, the cluster is stable for long days. Apart from vitro reconstitution, for understanding the cluster assembly in vivo in ORP, the ATCUN tag (Amino terminus Cu and Ni binding motif) [152] is inserted at N-terminus in the ORP (Ala1Ser2His3-native amino acid residues). The NMR spectrum of that event indicates that metal-cofactor formation has occurred through inter-molecular protein–protein interaction [30].

A series of metal derivatives replacing the native copper with iron, cobalt, nickel, and cadmium ions were reported. All derivatives, $[S_2MoS_2FeS_2MoS_2]^{3-}$ (ORP-Fe), $[S_2MoS_2CoS_2MoS_2]^{3-}$ (ORP-Co), $[S_2MoS_2NiS_2MoS_2]^{3-}$ (ORP-Ni) and $[S_2MoS_2CdS_2MoS_2]^{2-}$ (ORP-Cd), were synthesized using the apo-ORP as template [40]. These substitutions of the diamagnetic Cu^{I} ion ($[S_2MoS_2CuS_2MoS_2]^{3-}$) gave origin to either paramagnetic or NMR active derivatives. Therefore, the ORP-Fe cofactor is also interested as a minimal model of Fe-Mo-co in nitrogenase [13] and other derivatives such as $[S_2MoS_2FeS_2MoS_2]^{3-}$ (ORP-Fe), $[S_2MoS_2CoS_2MoS_2]^{3-}$ (ORP-Co), and $[S_2MoS_2NiS_2MoS_2]^{3-}$ (ORP-Ni) are EPR probe whereas $[S_2MoS_2CdS_2MoS_2]^{2-}$ (ORP-Cd) is NMR probe. This work represents the power of the protein matrix to determine the final shape and structure of metal-cofactor in situ formation. Moreover, host molecules encapsulate a variety of guests but the specific size and shape into the plastic cavity through non-covalent interactions.

Spectroscopic Probes ORP

The metal cofactor of ORP contains Cu^{I} and Mo^{VI}, which are EPR inactive metal species and have no satisfactory NMR reporters that can help to investigate the location of metal-cofactor in the protein cavity. The 3D structure obtained by NMR spectroscopy displays that the mixed metal cluster is located inside the ORP protein cavity by non-covalent interactions, including H-bondings, electrostatic and hydrophobic interactions with an array of amino acid residues [146,148]. However, the exact binding locations of the metal cofactor in apo-ORP remain elusive. The amino acids are crucial for stabilizing the mixed metal–sulfur center within the protein cavity, and their identification is crucial for understanding the structure and the function of the metalloproteins. Therefore, the biophysical characterization of the binding pocket within the protein is important and different types of spectroscopic tools are now available for a diagnostic. Therefore, the metal cofactor is manipulated by substituting the diamagnetic Cu^{I} atom with paramagnetic or interesting NMR active metal reporters or coordinating the metal core with a small organic thiol ligand (^{1}H or ^{19}F NMR). In order to achieve our aims, we have reported a few synthetic inorganic semi-model compounds, where Mo-Cu cluster coordinated nonbiological thiols and fluorinated thiols are characterized by ^{1}H-NMR and ^{19}F-NMR respectively [153,154]. In addition, reconstitution of the NMR active cadmium atom into $[S_2MoS_2CuS_2MoS_2]^{3-}$ is originated a ^{113}Cd-NMR active compound, $[S_2MoS_2CdS_2MoS_2]^{2-}$ [153]. The ^{19}F-NMR of Mo/Cu-thiol complexes shows a higher chemical shift, which is advantageous for the NMR probe as an alternative to ^{1}H-NMR, circumventing the narrow window of observation of protons and also the large number of protein-bearing protons that mask the protons in synthesized compounds. Thus, our synthetic strategy of synthesis was directed toward ^{19}F-NMR, because ^{19}F-NMR has no background signal, a broad window of chemical shifts (+400 to −400 ppm), and high sensitivity compared to ^{1}H-NMR. Moreover, because the chemical shift of ^{19}F-NMR is highly influenced by the local environment at the active site, the ^{19}F-NMR may be tuned by the vicinity of amino acids in proteins.

Transition metal ions (such as iron, cobalt, and nickel) have been used extensively as spin probes for the study of metalloproteins by means of NMR and EPR spectroscopy techniques. In this respect, the diamagnetic Cu^I ion in $[S_2MoS_2CuS_2MoS_2]^{3-}$ is replaced by a variety of paramagnetic metal ions ($M = Fe^{I/II}$, $Co^{I/II}$ and $Ni^{I/II}$) to obtain paramagnetic probes of ORP [40].

5. Conclusions

Enzymes are complex molecules that may or not contain metals at the catalytic site, where chemical transformations occur with amazing selectivity and at high rates. Of the known enzymes, one-third contain metals coordinated by the side chains of amino acids of the polypeptide chain and/or cofactors. In this case, the substrate is activated at the metal site.

Due to the chemical complexity of the system (large molecular mass, multiple subunit composition, and intricate architectural structures involving metals), the study of model compounds, retaining functional, structural (or both) characteristics has the advantage of working with a smaller size problem, more suitable for biophysical studies enabling to an inorganic chemistry approach for revealing the metal active site properties. Metalloenzymes use a wide range of metals in a variety of structural arrangements and geometries, most in parallel with inorganic compounds, but others are still a challenge for synthetic chemistry. Iron contained in iron–sulfur centers and in hemes are the most ubiquitous, but several other transition metals have specific roles, such as Ni, Mo, Cu, Zn, and others. Modeling efforts also represent an opportunity for further exploring new applications and functionalities.

The chemical design of models for metalloprotein active sites can be based on small inorganic compounds and now extend to peptides, protein-based synthetic analogs, and simple proteins that are used as templates (or scaffolds). As explained in this short review, we use as case studies three native scaffolds that provide very rich sulfur environments, used as templates for the synthesis of non-native metal clusters that can be mono, multi, and mixed metal–sulfur clusters, which mimic native metalloenzymes involved in key biological steps. Another advantage is the design of metal sites that can be quite useful as spectroscopic probes.

Author Contributions: Conceptualization, methodology, writing—original draft preparation, writing—review and editing, B.K.M. and J.J.G.M. All authors have read and agreed to the published version of the manuscript.

Funding: This research was funded by FCT/MCTES through projects grant number PTDC/BTA-BTA/0935/2020, UIDB/50006/2020, UIDP/50006/2020, and PTDC/BTA-BTA/0935/202.

Institutional Review Board Statement: Not applicable.

Informed Consent Statement: Not applicable.

Data Availability Statement: Not applicable.

Acknowledgments: BKM gives thanks to the Cluster University of Jammu, India. JJGM received financial support from PT national funds (FCT/MCTES) through projects PTDC/BTA-BTA/0935/2020, UIDB/50006/2020, UIDP/50006/2020, and PTDC/BTA-BTA/0935/2020.

Conflicts of Interest: The authors declare no conflict of interest.

Abbreviations

Rd	Rubredoxin
ORP	Orange Protein
Fd	Ferredoxin
CODH	Carbon monoxide dehydrogenase
cyt-c3	cytochrome c3
Pf	*Pyrococcus furiosus*
Dg	*Desulfovibrio gigas*
Dd	*Desulfovibrio desulfuricans*
Ch	*Carboxydothermus hydrogenoformans*
Rr	*Rhodospirillum rubrum*
Mt	*Moorella thermoacetica*
TTM	Tetrathiomolybdate
EXAFS	Extended X-ray absorption fine structure
PDB	Protein data bank
ET	Electron transfer
NHE	Normal hydrogen electrode
NMR	Nuclear magnetic resonance

References

1. Ragsdale, S.W. Metals and Their Scaffolds to Promote Difficult Enzymatic Reactions. *Chem. Rev.* **2006**, *106*, 3317–3337. [CrossRef] [PubMed]
2. Nastri, F.; D'Alonzo, D.; Leone, L.; Zambrano, G.; Pavone, V.; Lombardi, A. Engineering Metalloprotein Functions in Designed and Native Scaffolds. *Trends Biochem. Sci.* **2019**, *44*, 1022–1040. [CrossRef] [PubMed]
3. Bell, E.L.; Finnigan, W.; France, S.P.; Green, A.P.; Hayes, M.A.; Hepworth, L.J.; Lovelock, S.L.; Niikura, H.; Osuna, S.; Romero, E.; et al. Biocatalysis. *Nat. Rev. Methods Primers* **2021**, *1*, 46. [CrossRef]
4. Liu, J.; Chakraborty, S.; Hosseinzadeh, P.; Yu, Y.; Tian, S.; Petrik, I.; Ambika Bhagi, A.; Lu, Y. Metalloproteins Containing Cytochrome, Iron–Sulfur, or Copper Redox Centers. *Chem. Rev.* **2014**, *114*, 4366–4469. [CrossRef]
5. Valdez, C.E.; Smith, Q.A.; Nechay, M.R.; Alexandrova, A.N. Mysteries of Metals in Metalloenzymes. *Acc. Chem. Res.* **2014**, *47*, 3110–3117. [CrossRef] [PubMed]
6. Mirts, E.N.; Bhagi-Damodaran, A.; Lu, Y. Understanding and Modulating Metalloenzymes with Unnatural Amino Acids, Non-Native Metal Ions, and Non-Native Metallocofactors. *Acc. Chem. Res.* **2019**, *52*, 935–944. [CrossRef] [PubMed]
7. Maiti, B.K.; Maia, L.B.; Moura, J.J.G. Sulfide and transition metals-A partnership for life. *J. Inorg. Biochem.* **2022**, *227*, 111687. [CrossRef]
8. Solomon, E.I.; Heppner, D.E.; Johnston, E.M.; Ginsbach, J.W.; Cirera, J.; Qayyum, M.; Kieber-Emmons, M.T.; Kjaergaard, C.H.; Hadt, R.G.; Tian, L. Copper Active Sites in Biology. *Chem. Rev.* **2014**, *114*, 3659–3853. [CrossRef]
9. Maiti, B.K.; Almeida, R.M.; Moura, I.; Moura, J.J.G. Rubredoxins derivatives: Simple sulphur-rich coordination metal sites and its relevance for biology and chemistry. *Coord. Chem. Rev.* **2017**, *352*, 379–397. [CrossRef]
10. Denisov, I.G.; Makris, T.M.; Sligar, S.G.; Schlichting, I. Structure and Chemistry of Cytochrome P450. *Chem. Rev.* **2005**, *105*, 2253–2278. [CrossRef] [PubMed]
11. Chen, Q.; Yuan, G.; Yuan, T.; Zeng, H.; Zou, Z.-R.; Tu, Z.-C.; Gao, J.; Zou, Y. Set of Cytochrome P450s Cooperatively Catalyzes the Synthesis of a Highly Oxidized and Rearranged Diterpene-Class Sordarinane Architecture. *J. Am. Chem. Soc.* **2022**, *144*, 3580–3589. [CrossRef]
12. Hille, R.; Hall, J.; Basu, P. The Mononuclear Molybdenum Enzymes. *Chem. Rev.* **2014**, *114*, 3963–4038. [CrossRef]
13. Einsle, O.; Rees, D.C. Structural Enzymology of Nitrogenase Enzymes. *Chem. Rev.* **2020**, *120*, 4969–5004. [CrossRef] [PubMed]
14. Seefeldt, L.C.; Yang, Z.-Y.; Lukoyanov, D.A.; Harris, D.F.; Dean, D.R.; Raugei, S.; Hoffman, B.M. Reduction of Substrates by Nitrogenases. *Chem. Rev.* **2020**, *120*, 5082–5106. [CrossRef]
15. McEvoy, J.P.; Brudvig, G.W. Water-splitting chemistry of photosystem II. *Chem. Rev.* **2006**, *106*, 4455–4483. [CrossRef] [PubMed]
16. Marchiori, D.A.; Debus, R.J.; Britt, R.D. Pulse EPR Spectroscopic Characterization of the S 3 State of the Oxygen-Evolving Complex of Photosystem II Isolated from Synechocystis. *Biochemistry* **2020**, *59*, 4864–4872. [CrossRef] [PubMed]
17. Lubitz, W.; Ogata, H.; Rüdiger, O.; Reijerse, E. Hydrogenases. *Chem. Rev.* **2014**, *114*, 4081–4148. [CrossRef]
18. Tai, H.; Hirota, S.; Stripp, S.T. Proton Transfer Mechanisms in Bimetallic Hydrogenases. *Acc. Chem. Res.* **2021**, *54*, 232–241. [CrossRef]
19. Cohen, S.E.; Can, M.; Wittenborn, E.C.; Hendrickson, R.A.; Ragsdale, S.W.; Drennan, C.L. Crystallographic Characterization of the Carbonylated A-Cluster in Carbon Monoxide Dehydrogenase/Acetyl-CoA Synthase. *ACS Catal.* **2020**, *10*, 9741–9746. [CrossRef]
20. Reginald, S.S.; Etzerodt, M.; Fapyane, D.; Chang, I.S. Functional Expression of a Mo–Cu-Dependent Carbon Monoxide Dehydrogenase (CODH) and Its Use as a Dissolved CO Bio-microsensor. *ACS Sens.* **2021**, *6*, 2772–2782. [CrossRef] [PubMed]
21. Kazlauskas, R.J. Enhancing catalytic promiscuity for biocatalysis. *Curr. Opin. Chem. Biol.* **2005**, *9*, 195–201. [CrossRef] [PubMed]

22. Khersonsky, O.; Roodveldt, C.; Tawfik, D.S. Enzyme promiscuity: Evolutionary and mechanistic aspects. *Curr. Opin. Chem. Biol.* **2006**, *10*, 498–508. [CrossRef] [PubMed]

23. Toscano, M.D.; Woycechowsky, K.J.; Hilvert, D. Minimalist active-site redesign: Teaching old enzymes new tricks. *Angew. Chem. Int. Ed.* **2007**, *46*, 3212–3236. [CrossRef]

24. Fernandez-Gacio, A.; Codina, A.; Fastrez, J.; Riant, O.; Soumillion, P. Transforming Carbonic Anhydrase into Epoxide Synthase by Metal Exchange. *ChemBioChem* **2006**, *7*, 1013–1016. [CrossRef] [PubMed]

25. Lu, Y.; Berry, S.M.; Pfister, T.D. Engineering Novel Metalloproteins: Design of Metal-Binding Sites into Native Protein Scaffolds. *Chem. Rev.* **2001**, *101*, 3047–3080. [CrossRef]

26. Yu, F.; Cangelosi, V.M.; Zastrow, M.L.; Tegoni, M.; Plegaria, J.S.; Tebo, A.G.; Mocny, C.S.; Ruckthong, L.; Qayyum, H.; Pecoraro, V.L. Protein Design: Toward Functional Metalloenzymes. *Chem. Rev.* **2014**, *114*, 3495–3578. [CrossRef] [PubMed]

27. Beinert, H.; Kennedy, M.C.; Stout, C.D. Aconitase as Ironminus signSulfur Protein, Enzyme, and Iron-Regulatory Protein. *Chem. Rev.* **1996**, *96*, 2335–2374. [CrossRef] [PubMed]

28. Holm, R.H.; Lo, W. Structural Conversions of Synthetic and Protein-Bound Iron–Sulfur Clusters. *Chem. Rev.* **2016**, *116*, 13685–13713. [CrossRef] [PubMed]

29. George, G.N.; Pickering, I.J.; Yu, E.Y.; Prince, R.C.; Bursakov, S.A.; Gavel, O.Y.; Moura, I.; Moura, J.J.G. A Novel Protein-Bound Copper–Molybdenum Cluster. *J. Am. Chem. Soc.* **2000**, *122*, 8321–8322. [CrossRef]

30. Maiti, B.K.; Almedia, R.; Maia, L.B.; Moura, I.; Moura, J.J.G. Insights into the Molybdenum/Copper Heterometallic Cluster Assembly in the Orange Protein: Probing Intermolecular Interactions with an Artificial Metal-Binding ATCUN Tag. *Inorg Chem.* **2017**, *56*, 8900–8911. [CrossRef]

31. Coleman, J.E. Cadmium-113 nuclear magnetic resonance applied to metalloproteins. *Methods Enzymol.* **1993**, *227*, 16–43. [PubMed]

32. Maret, W.; Vallee, B.L. Cobalt as probe and label of proteins. *Methods Enzymol.* **1993**, *226*, 52–71.

33. Münck, E.; Ksurerus, K.; Hendrich, M.P. Combining Mossbauer spectroscopy with integer spin electron paramagnetic resonance. *Methods Enzymol.* **1993**, *227*, 463–479. [PubMed]

34. Moura, J.J.; Macedo, A.L.; Nuno Palma, P. Ferredoxins. *Methods Enzymol.* **1994**, *243*, 165–188. [PubMed]

35. Butt, J.N.; Fawcett, S.E.J.; Breton, J.; Thomson, J.A.; Armstrong, F.A. Electrochemical Potential and pH Dependences of [3Fe-4S] ↔ [M3Fe-4S] Cluster Transformations (M = Fe, Zn, Co, and Cd) in Ferredoxin III from Desulfovibrio africanus and Detection of a Cluster with M = Pb. *J. Am. Chem. Soc.* **1997**, *119*, 9729–9737. [CrossRef]

36. Conover, R.C.; Park, J.B.; Adams, M.W.W.; Johnson, M.K. Formation and properties of an iron-nickel sulfide (NiFe3S4) cluster in Pyrococcus furiosus ferredoxin. *J. Am. Chem. Soc.* **1990**, *112*, 4562–4564. [CrossRef]

37. Slater, J.W.; Marguet, S.C.; Monaco, H.A.; Shafaat, H.S. Going beyond Structure: Nickel-Substituted Rubredoxin as a Mechanistic Model for the [NiFe] Hydrogenases. *J. Am. Chem. Soc.* **2018**, *140*, 10250–10262. [CrossRef]

38. Saint-Martin, P.; Lespinat, P.A.; Fauque, G.; Berlier, Y.; Legall, J.; Moura, I.; Teixeira, M.; Xavier, A.V.; Moura, J.J. Hydrogen production and deuterium-proton exchange reactions catalyzed by Desulfovibrio nickel(II)-substituted rubredoxins. *Proc. Natl. Acad. Sci. USA* **1988**, *85*, 9378–9380. [CrossRef] [PubMed]

39. Lewis, L.C.; Shafaat, H.S. Reversible Electron Transfer and Substrate Binding Support [NiFe3S4] Ferredoxin as a Protein-Based Model for [NiFe] Carbon Monoxide Dehydrogenase. *Inorg. Chem.* **2021**, *60*, 13869–13875. [CrossRef]

40. Maiti, B.K.; Maia, L.B.; Pauleta, S.R.; Moura, I.; Moura, J.J.G. Protein-Assisted Formation of Molybdenum Heterometallic Clusters: Evidence for the Formation of S_2MoS_2–M–S_2MoS_2 Clusters with M = Fe, Co, Ni, Cu, or Cd within the Orange Protein. *Inorg. Chem.* **2017**, *56*, 2210–2220. [CrossRef]

41. Rivas, M.G.; Carepo, M.S.P.; Mota, C.S.; Korbas, M.; Durand, M.-C.; Lopes, A.T.; Brondino, C.D.; Pereira, A.S.; George, G.N.; Dolla, A.; et al. Molybdenum induces the expression of a protein containing a new heterometallic Mo-Fe cluster in Desulfovibrio alaskensis. *Biochemistry* **2009**, *48*, 873–882. [CrossRef] [PubMed]

42. Maiti, B.K.; Pal, K.; Sarkar, S. Plasticity in $[(R_{4-x}R^1_x)_4N]_4[Cu_4\{S_2C_2(CN)_2\}_4]$ (x = 0–4) is Molded by a Guest Cation on an Elastic Anionic Host. *Eur. J. Inorg. Chem.* **2008**, *2008*, 2407–2420. [CrossRef]

43. Lehn, J.-M. *Supramolecular Chemistry: Concepts and Perspectives*; VCH: Weinheim, Germany, 1995.

44. Spiro, T.G. *Iron-Sulfur Proteins: Metal Ions in Biology*; Thomas, G., Spiro, Eds.; Wiley-Interscience: New York, NY, USA, 1982; Volume IV.

45. Sofia, R.; Pauleta, S.R.; Grazina, R.; Carepo, M.S.P.; Moura, J.J.M.; Moura, I. Iron-Sulfur Clusters–Functions of an Ancient Metal Site, Comprehensive Inorganic Chemistry III from Biology to Nanotechnology. In *Bioinorganic Chemistry and Homogeneous Biomimetic Inorganic Catalysis*; Pecoraro, V., Guo, Z., Eds.; Elsevier: Amsterdam, The Netherlands, 2022; Volume 2, ISBN 9780081026885. *in press*.

46. Sykes, A.G.; Cammack, R. Iron-Sulfur Proteins. In *Advances in Inorganic Chemistry*; Academic Press: San Diego, CA, USA, 1999; Volume 47.

47. Maiti, B.K.; Moura, I.; Moura, J.J.G.; Pauleta, S.R. The small iron-sulfur protein from the ORP operon binds a [2Fe-2S] cluster. *Biochim. Biophys. Acta (BBA)–Bioenerg.* **2016**, *1857*, 1422–1429. [CrossRef] [PubMed]

48. Tavares, P.; Pereira, A.S.; Krebs, C.; Ravi, N.; Moura, J.J.; Moura, I.; Huynh, B.H. Spectroscopic characterization of a novel tetranuclear Fe cluster in an iron-sulfur protein isolated from Desulfovibrio desulfuricans. *Biochemistry* **1998**, *37*, 2830–2842. [CrossRef]

49. Maiti, B.K. Cross-talk Between (Hydrogen)Sulfite and Metalloproteins: Impact on Human Health. *Chem.-A Eur. J.* **2022**, *28*, e202104342. [CrossRef]

50. Reed, C.J.; Lam, Q.N.; Mirts, E.N.; Lu, Y. Molecular understanding of heteronuclear active sites in heme–copper oxidases, nitric oxide reductases, and sulfite reductases through biomimetic modelling. *Chem. Soc. Rev.* **2021**, *50*, 2486–2539. [CrossRef]

51. Moura, I.; Pereira, A.S.; Tavares, P.; Moura, J.J.G. Simple and Complex Iron-Sulfur Proteins in Sulfate Reducing Bacteria. *Adv. Inorg. Chem.* **1999**, *47*, 361–419.

52. Jeoung, J.-H.; Martins, B.M.; Dobbek, H. Double-Cubane [8Fe9S] Clusters: A Novel Nitrogenase-Related Cofactor in Biology. *Chembiochem* **2020**, *21*, 1710–1716. [CrossRef]

53. Del Barrio, M.; Sensi, M.; Fradale, L.; Bruschi, M.; Greco, C.; de Gioia, L.; Bertini, L.; Fourmond, V.; Léger, C. Interaction of the H-Cluster of FeFe Hydrogenase with Halides. *J. Am. Chem. Soc.* **2018**, *140*, 5485–5492. [CrossRef] [PubMed]

54. Volbeda, A.; Fontecilla-Camps, J.C. Structural bases for the catalytic mechanism of Ni-containing carbon monoxide dehydrogenases. *Dalton Trans.* **2005**, *21*, 3443–3450. [CrossRef]

55. Jeoung, J.H.; Dobbek, H. ATP-dependent substrate reduction at an [Fe8S9] double-cubane cluster. *Proc. Natl. Acad. Sci. USA* **2018**, *115*, 2994–2999. [CrossRef] [PubMed]

56. Wagner, T.; Koch, J.; Ermler, U.; Shima, S. Methanogenic heterodisulfide reductase (HdrABC-MvhAGD) uses two noncubane [4Fe-4S] clusters for reduction. *Science* **2017**, *357*, 699–703. [CrossRef]

57. Vervoort, J.; Heering, D.; Peelen, S.; van Berkel, W. Flavodoxins. *Methods Enzymol.* **1994**, *243*, 188–203. [PubMed]

58. Wastl, J.; Sticht, H.; Maier, U.-G.; Rösch, P.; Hoffmann, S. Identification and characterization of a eukaryotically encoded rubredoxin in a cryptomonad alga. *FEBS Lett.* **2000**, *471*, 191–196. [CrossRef]

59. Zauner, S.; Fraunholz, M.; Wastl, J.; Penny, S.; Beaton, M.; Cavalier-Smith, T.; Maier, U.G.; Douglas, S. Chloroplast protein and centrosomal genes, a tRNA intron, and odd telomeres in an unusually compact eukaryotic genome, the cryptomonad nucleomorph. *Proc. Natl. Acad. Sci. USA* **2000**, *97*, 200–205. [CrossRef]

60. Moura, I.; Bruschi, M.; Legall, J.; Moura, J.J.G.; Xavier, A.V. Isolation and characterization of desulforedoxin, a new type of non-heme iron protein from Desulfovibrio gigas. *Biochem. Biophys. Res. Commun.* **1977**, *75*, 1037–1044. [CrossRef]

61. Zanello, P. The competition between chemistry and biology in assembling iron–sulfur derivatives. Molecular structures and electrochemistry. Part I. {Fe(SγCys)4} proteins. *Coord. Chem. Rev.* **2013**, *257*, 1777–1805. [CrossRef]

62. Hagelueken, G.; Wiehlmann, L.; Adams, T.M.; Kolmar, H.; Heinz, D.W.; Tümmler, B.; Schubert, W.D. Crystal structure of the electron transfer complex rubredoxin rubredoxin reductase of Pseudomonas aeruginosa. *Proc. Natl. Acad. Sci. USA* **2007**, *104*, 12276–12281. [CrossRef] [PubMed]

63. Yoon, K.S.; Hille, R.; Hemann, C.; Tabita, F.R. Rubredoxin from the green sulfur bacterium Chlorobium tepidum functions as an electron acceptor for pyruvate ferredoxin oxidoreductase. *J. Biol. Chem.* **1999**, *274*, 29772–29778. [CrossRef]

64. Ragsdale, S.W.; Ljungdahl, L.G.; DerVartanian, D.V. Isolation of carbon monoxide dehydrogenase from Acetobacterium woodii and comparison of its properties with those of the Clostridium thermoaceticum enzyme. *J. Bacteriol.* **1983**, *155*, 1224–1237. [CrossRef]

65. Lin, I.J.; Gebel, E.B.; Machonkin, T.E.; Westler, W.M.; Markley, J.L. Changes in hydrogen-bond strengths explain reduction potentials in 10 rubredoxin variants. *Proc. Natl. Acad. Sci. USA* **2005**, *102*, 14581–14586. [CrossRef]

66. Sun, N.; Dey, A.; Xiao, Z.; Wedd, A.G.; Hodgson, K.O.; Hedman, B.; Solomon, E.I. Solvation effects on S K-edge XAS spectra of Fe-S proteins: Normal and inverse effects on WT and mutant rubredoxin. *J. Am. Chem. Soc.* **2010**, *132*, 12639–12647. [CrossRef]

67. Min, T.; Ergenekan, C.E.; Eidsness, M.K.; Ichiye, T.; Kang, C. Leucine 41 is a gate for water entry in the reduction of Clostridium pasteurianum rubredoxin. *Protein Sci.* **2001**, *10*, 613–621. [CrossRef]

68. Jenney, F.E.; Adams, M.W.W. Rubredoxin from *Pyrococcus furiosus*. In *Methods in Enzymology*; Elsevier: Amsterdam, The Netherlands, 2001; Volume 334, pp. 45–55.

69. Moura, I.; Teixeira, M.; LeGall, J.; Moura, J.J. Spectroscopic studies of cobalt and nickel substituted rubredoxin and desulforedoxin. *J. Inorg. Biochem.* **1991**, *44*, 127–139. [CrossRef]

70. Xx Thapper, A.; Rizzi, A.C.; Brondino, C.D.; Wedd, A.G.; Pais, R.J.; Maiti, B.K.; Moura, I.; Pauleta, S.R.; Moura, J.J. Copper-substituted forms of the wild type and C42A variant of rubredoxin. *J. Inorg. Biochem.* **2013**, *127*, 232–237. [CrossRef]

71. Maiti, B.K.; Maia, L.B.; Moro, A.J.; Lima, J.C.; Cordas, C.M.; Moura, I.; Moura, J.J.G. Unusual Reduction Mechanism of Copper in Cysteine-Rich Environment. *Inorg. Chem.* **2018**, *57*, 8078–8088. [CrossRef] [PubMed]

72. Maiti, B.K.; Maia, L.B.; Silveira, C.M.; Todorovic, S.; Carreira, C.; Carepo, M.S.; Grazina, R.; Moura, I.; Pauleta, S.R.; Moura, J.J.G. Incorporation of molybdenum in rubredoxin: Models for mononuclear molybdenum enzymes. *J. Biol. Inorg. Chem.* **2015**, *20*, 821–829. [CrossRef] [PubMed]

73. Slater, J.W.; Marguet, S.C.; Gray, M.E.; Monaco, H.A.; Sotomayor, M.; Shafaat, H.S. Power of the Secondary Sphere: Modulating Hydrogenase Activity in Nickel-Substituted Rubredoxin. *ACS Catal.* **2019**, *9*, 8928–8942. [CrossRef]

74. Shafaat, H.S.; Rüdiger, O.; Ogata, H.; Lubitz, W. [NiFe] hydrogenases: A common active site for hydrogen metabolism under diverse conditions. *Biochim. Biophys. Acta BBA-Bioenerg.* **2013**, *1827*, 986–1002. [CrossRef] [PubMed]

75. Volbeda, A.; Charon, M.H.; Piras, C.; Hatchikian, E.C.; Frey, M.; Fontecilla-Camps, J.C. Crystal structure of the nickel-iron hydrogenase from Desulfovibrio gigas. *Nature* **1995**, *373*, 580–587. [CrossRef] [PubMed]

76. Volbeda, A.; Garcin, E.; Piras, C.; de Lacey, A.L.; Fernandez, V.M.; Hatchikian, E.C.; Frey, M.; Fontecilla-Camps, J.C. Structure of the [NiFe] Hydrogenase Active Site: Evidence for Biologically Uncommon Fe Ligands. *J. Am. Chem. Soc.* **1996**, *118*, 12989–12996. [CrossRef]

77. Ogata, H.; Nishikawa, K.; Lubitz, W. Hydrogens detected by subatomic resolution protein crystallography in a [NiFe] hydrogenase. *Nature* **2015**, *520*, 571–574. [CrossRef] [PubMed]

78. Gloaguen, F.; Rauchfuss, T.B. Small molecule mimics of hydrogenases: Hydrides and redox. *Chem. Soc. Rev.* **2009**, *38*, 100–108. [CrossRef] [PubMed]

79. Roy, A.; Madden, C.; Ghirlanda, G. Photo-induced hydrogen production in a helical peptide incorporating a [FeFe] hydrogenase active site mimic. *Chem. Commun.* **2012**, *79*, 9816–9818. [CrossRef] [PubMed]

80. Utschig, L.M.; Silver, S.C.; Mulfort, K.L.; Tiede, D.M. Nature-driven photochemistry for catalytic solar hydrogen production: A Photosystem I-transition metal catalyst hybrid. *J. Am. Chem. Soc.* **2011**, *133*, 16334–16337. [CrossRef] [PubMed]

81. Thoi, V.S.; Sun, Y.; Long, J.R.; Chang, C.J. Complexes of earth-abundant metals for catalytic electrochemical hydrogen generation under aqueous conditions. *Chem. Soc. Rev.* **2013**, *42*, 2388–2400. [CrossRef] [PubMed]

82. Schilter, D.; Camara, J.M.; Huynh, M.T.; Hammes-Schiffer, S.; Rauchfuss, T.B. Hydrogenase Enzymes and Their Synthetic Models: The Role of Metal Hydrides. *Chem. Rev.* **2016**, *116*, 8693–8749. [CrossRef]

83. Denny, J.A.; Darensbourg, M.Y. Metallodithiolates as ligands in coordination, bioinorganic, and organometallic chemistry. *Chem. Rev.* **2015**, *115*, 5248–5273. [CrossRef]

84. Ogo, S.; Ichikawa, K.; Kishima, T.; Matsumoto, T.; Nakai, H.; Kusaka, K.; Ohhara, T. A functional [NiFe]hydrogenase mimic that catalyzes electron and hydride transfer from H_2. *Science* **2013**, *339*, 682–684. [CrossRef] [PubMed]

85. Slater, J.W.; Shafaat, H.S. Nickel-Substituted Rubredoxin as a Minimal Enzyme Model for Hydrogenase. *J. Phys. Chem. Lett.* **2015**, *6*, 3731–3736. [CrossRef]

86. Slater, J.W.; Marguet, S.C.; Cirino, S.L.; Maugeri, P.T.; Shafaat, H.S. Experimental and DFT Investigations Reveal the Influence of the Outer Coordination Sphere on the Vibrational Spectra of Nickel-Substituted Rubredoxin, a Model Hydrogenase Enzyme. *Inorg. Chem.* **2017**, *56*, 3926–3938. [CrossRef]

87. Stevenson, M.J.; Marguet, S.C.; Schneider, C.R.; Shafaat, H.S. Light-Driven Hydrogen Evolution by Nickel-Substituted Rubredoxin. *Chem. Sustain. Chem.* **2017**, *10*, 4424–4429. [CrossRef] [PubMed]

88. Treviño, R.E.; Slater, J.W.; Shafaat, H.S. Robust Carbon-Based Electrodes for Hydrogen Evolution through Site-Selective Covalent Attachment of an Artificial Metalloenzyme. *ACS Appl. Energy Mater.* **2020**, *3*, 11099–11112.

89. Naughton, K.J.; Treviño, R.E.; Moore, P.J.; Wertz, A.E.; Dickson, J.A.; Shafaat, H.S. In Vivo Assembly of a Genetically Encoded Artificial Metalloenzyme for Hydrogen Production. *ACS Synth. Biol.* **2021**, *10*, 2116–2120. [CrossRef]

90. Detz, R.J.; Sakai, K.; Spiccia, L.; Brudvig, G.W.; Sun, L.; Reek, J.N.H. Towards a Bioinspired-Systems Approach for Solar Fuel Devices. *Chem. Plus Chem.* **2016**, *81*, 1024–1027. [CrossRef] [PubMed]

91. Utschig, L.M.; Soltau, S.R.; Tiede, D.M. Light-driven hydrogen production from Photosystem I-catalyst hybrids. *Curr. Opin. Chem. Biol.* **2015**, *25*, 1–8. [CrossRef] [PubMed]

92. Nichols, E.M.; Gallagher, J.J.; Liu, C.; Su, Y.; Resasco, J.; Yu, Y.; Sun, Y.; Yang, P.; Chang, M.C.; Chang, C.J. Hybrid bioinorganic approach to solar-to-chemical conversion. *Proc. Natl. Acad. Sci. USA* **2015**, *112*, 11461–11466. [CrossRef] [PubMed]

93. Reisner, E. Solar Hydrogen Evolution with Hydrogenases: From Natural to Hybrid Systems. *Eur. J. Inorg. Chem.* **2011**, *2011*, 1005–1016. [CrossRef]

94. Marguet, S.C.; Stevenson, M.J.; Shafaat, H.S. Intramolecular Electron Transfer Governs Photoinduced Hydrogen Evolution by Nickel-Substituted Rubredoxin: Resolving Elementary Steps in Solar Fuel Generation. *J. Phys. Chem. B* **2019**, *123*, 9792–9800. [CrossRef] [PubMed]

95. Yoo, S.J.; Meyer, J.; Achim, C.; Peterson, J.; Hendrich, M.P.; Münck, E. Mössbauer, EPR, and MCD studies of the C9S and C42S variants of *Clostridium pasteurianum* rubredoxin and MDC studies of the wild-type protein. *J. Biol. Inorg. Chem.* **2000**, *5*, 475–487. [CrossRef]

96. Moura, I.; Tavares, P.; Moura, J.J.; Ravi, N.; Huynh, B.H.; Liu, M.Y.; LeGall, J. Purification and characterization of desulfoferrodoxin. A novel protein from *Desulfovibrio desulfuricans* (ATCC 27774) and from *Desulfovibrio vulgaris* (strain Hildenborough) that contains a distorted rubredoxin center and a mononuclear ferrous center. *J. Biol. Chem.* **1990**, *265*, 21596–21602. [CrossRef]

97. Werth, M.T.; Kurtz, D.M., Jr.; Moura, I.; LeGall, J. Proton NMR spectra of rubredoxins: New resonances assignable to .alpha.-CH and beta-CH_2 hydrogens of cysteinate ligands to iron(II). *J. Am. Chem. Soc.* **1987**, *109*, 273–275. [CrossRef]

98. Almeida, R.M.; Pauleta, S.R.; Moura, I.; Moura, J.J. Rubredoxin as a paramagnetic relaxation-inducing probe. *J. Inorg. Biochem.* **2009**, *103*, 1245–1253. [CrossRef] [PubMed]

99. Almeida, R.; Turano, P.; Moura, I.M.A.M.G.D.; Pauleta, S.R.; Moura, J.J.G.D. Superoxide Reductase: Different Interaction Modes with its Two Redox Partners. *Chem. Bio Chem.* **2013**, *14*, 1858–1866. [CrossRef]

100. Lamosa, P.; Brennan, L.; Vis, H.; Turner, D.L.; Santos, H. NMR structure of Desulfovibrio gigas rubredoxin: A model for studying protein stabilization by compatible solutes. *Extremophiles* **2001**, *5*, 303–311. [CrossRef] [PubMed]

101. Goodfellow, B.J.; Nunes, S.G.; Rusnak, F.; Moura, I.; Ascenso, C.; Moura, J.J.; Volkman, B.F.; Markley, J.L. Zinc-substituted *Desulfovibrio gigas* desulforedoxins: Resolving subunit degeneracy with nonsymmetric pseudocontact shifts. *Protein Sci.* **2002**, *11*, 2464–2470. [CrossRef] [PubMed]

102. Armitage, I.M.; Drakenberg, T.; Reilly, B. Use of ^{113}Cd NMR to probe the native metal binding sites in metalloproteins: An overview. *Met. Ions. Life. Sci.* **2013**, *11*, 117–144. [PubMed]

103. Vasak, M. Application of ^{113}Cd NMR to metallothioneins. *Biodegradation* **1998**, *9*, 501–512. [CrossRef] [PubMed]

104. Summers, M.F. ^{113}Cd NMR spectroscopy of coordination compounds and proteins. *Coord. Chem. Rev.* **1988**, *86*, 43–134. [CrossRef]

105. Ayhan, M.; Xiao, Z.; Lavery, M.J.; Hamer, A.M.; Nugent, K.W.; Scrofani, S.D.B.; Guss, M.; Wedd, A.G. The Rubredoxin from Clostridium pasteurianum: Mutation of the Conserved Glycine Residues 10 and 43 to Alanine and Valine. *Inorg. Chem.* **1996**, *35*, 5902–5911. [CrossRef]

106. Fukuyama, K.; Matsubara, H.; Tsukihara, T.; Katsube, Y. Structure of [4Fe-4S] ferredoxin from Bacillus thermoproteolyticus refined at 2.3 A resolution. Structural comparisons of bacterial ferredoxins. *J. Mol. Biol.* **1989**, *210*, 383–398. [CrossRef]

107. Fukuyama, K. *Handbook of Metalloproteins*; Messerschmidt, A., Huber, R., Wieghardt, K., Poulos, T., Eds.; Wiley: New York, NY, USA, 2001; pp. 543–552.

108. Johnson, D.C.; Dean, D.R.; Smith, A.D.; Johnson, M.K. Structure, function, and formation of biological iron-sulfur clusters. *Annu. Rev. Biochem.* **2005**, *74*, 247–281. [CrossRef] [PubMed]

109. Py, B.; Barras, F. Building Fe-S proteins: Bacterial strategies. *Nat. Rev. Microbiol.* **2010**, *8*, 436–446. [CrossRef] [PubMed]

110. Mettert, E.L.; Kiley, P.J. Fe-S proteins that regulate gene expression. *Biochim. Biophys. Acta (BBA)-Mol. Cell Res.* **2015**, *1853*, 1284–1293. [CrossRef] [PubMed]

111. Johnson, M.K. *Encyclopedia of Inorganic Chemistry*; King, R.B., Ed.; Wiley: Oxford, UK, 1994; Volume 4, pp. 1896–1915.

112. Cammack, R. Iron—Sulfur Clusters in Enzymes: Themes and Variations. *Adv. Inorg. Chem.* **1992**, *38*, 281–322.

113. Duff, J.L.C.; Breton, J.L.J.; Butt, J.N.; Armstrong, F.A.; Thomson, A.J. Novel Redox Chemistry of [3Fe-4S] Clusters: Electrochemical Characterization of the All-Fe(II) Form of the [3Fe-4S] Cluster Generated Reversibly in Various Proteins and Its Spectroscopic Investigation in Sulfolobus acidocaldarius Ferredoxin. *J. Am. Chem. Soc.* **1996**, *118*, 8593–8603. [CrossRef]

114. Smith, E.T.; Blamey, J.M.; Zhou, Z.H.; Adams, M.W.W. A Variable-Temperature Direct Electrochemical Study of Metalloproteins from Hyperthermophilic Microorganisms Involved in Hydrogen Production from Pyruvate. *Biochemistry* **1995**, *34*, 7161–7169. [CrossRef] [PubMed]

115. Moreno, C.; Macedo, A.L.; Moura, I.; LeGall, J.; Moura, J.J. Redox properties of Desulfovibrio gigas [Fe$_3$S$_4$] and [Fe$_4$S$_4$] ferredoxins and heterometal cubane-type clusters formed within the [Fe$_3$S$_4$] core. Square wave voltammetric studies. *J. Inorg. Biochem.* **1994**, *53*, 219–234. [CrossRef]

116. Tong, J.; Feinberg, B.A. Direct square-wave voltammetry of superoxidized [4Fe-4S]$^{3+}$ aconitase and associated 3Fe/4Fe cluster interconversions. *J. Biol. Chem.* **1994**, *269*, 24920–24927. [CrossRef]

117. Shen, B.; Martin, L.L.; Butt, J.N.; Armstrong, F.A.; Stout, C.D.; Jensen, G.M.; Stephens, P.J.; La Mar, G.N.; Gorst, C.M.; Burgess, B.K. Azotobacter vinelandii ferredoxin I. Aspartate 15 facilitates proton transfer to the reduced [3Fe-4S] cluster. *J. Biol. Chem.* **1993**, *268*, 25928–25939. [CrossRef]

118. Butt, J.N.; Armstrong, F.A.; Breton, J.; George, S.J.; Thomson, A.J.; Hatchikian, E.C. Investigation of metal ion uptake reactivities of [3Fe-4S] clusters in proteins: Voltammetry of co-adsorbed ferredoxin-aminocyclitol films at graphite electrodes and spectroscopic identification of transformed clusters. *J. Am. Chem. Soc.* **1991**, *113*, 6663–6670. [CrossRef]

119. Shirakawa, T.; Takahashi, Y.; Wada, K.; Hirota, J.; Takao, T.; Ohmori, D.; Fukuyama, K. Identification of variant molecules of *Bacillus thermoproteolyticus* ferredoxin: Crystal structure reveals bound coenzyme A and an unexpected [3Fe-4S] cluster associated with a canonical [4Fe-4S] ligand motif. *Biochemistry* **2005**, *44*, 12402–12410. [CrossRef] [PubMed]

120. Kennedy, M.C.; Kent, T.A.; Emptage, M.; Merkle, H.; Beinert, H.; Münck, E. Evidence for the formation of a linear [3Fe-4S] cluster in partially unfolded aconitase. *J. Biol. Chem.* **1984**, *259*, 14463–14471. [CrossRef]

121. Conover, R.C.; Kowal, A.T.; Fu, W.G.; Park, J.B.; Aono, S.; Adams, M.W.; Johnson, M.K. Spectroscopic characterization of the novel iron-sulfur cluster in *Pyrococcus furiosus* ferredoxin. *J. Biol. Chem.* **1990**, *265*, 8533–8541. [CrossRef]

122. Busch, J.L.H.; Breton, J.L.; Bartlett, B.M.; Armstrong, F.A.; James, R.; Andrew, J.; Thomson, A.J. [3Fe-$_4$S]↔[4Fe-$_4$S] cluster interconversion in *Desulfovibrio africanus* ferredoxin III: Properties of an Asp14→Cys mutant. *Biochem. J.* **1997**, *323*, 95–102. [CrossRef]

123. Werst, M.M.; Kennedy, M.C.; Beinert, H.; Hoffman, B.M. Oxygen-17, proton, and deuterium electron nuclear double resonance characterization of solvent, substrate, and inhibitor binding to the iron-sulfur [$_4$Fe-$_4$S]+ cluster of aconitase. *Biochemistry* **1990**, *29*, 10526–10532. [CrossRef] [PubMed]

124. Calzolai, L.; Gorst, C.M.; Zhao, Z.-H.; Teng, Q.; Adams, M.W.W.; Mar, G.N.L. 1H NMR Investigation of the Electronic and Molecular Structure of the Four-Iron Cluster Ferredoxin from the Hyperthermophile Pyrococcus furiosus. Identification of Asp 14 as a Cluster Ligand in Each of the Four Redox States. *Biochemistry* **1995**, *34*, 11373–11384. [CrossRef]

125. Moura, I.; Moura, J.J.G.; Munck, E.; Papaefthymiou, V.; LeGall, J. Evidence for the formation of a cobalt-iron-sulfur (CoFe3S4) cluster in Desulfovibrio gigas ferredoxin II. *J. Am. Chem. Soc.* **1986**, *108*, 349–351. [CrossRef]

126. Fu, W.; Telser, J.; Hoffman, B.M.; Smith, E.T.; Adams, M.W.W.; Johnson, M.K. Interaction of Tl+ and Cs+ with the [Fe$_3$S$_4$] Cluster of Pyrococcus furiosus Ferredoxin: Investigation by Resonance Raman, MCD, EPR, and ENDOR Spectroscopy. *J. Am. Chem. Soc.* **1994**, *116*, 5722–5729. [CrossRef]

127. Finnegan, M.G.; Conover, R.C.; Park, J.-B.; Zhou, Z.H.; Adams, M.W.W.; Johnson, M.K. Electronic, Magnetic, Redox, and Ligand-Binding Properties of [MFe$_3$S$_4$] Clusters (M = Zn, Co, Mn) in *Pyrococcus furiosus* Ferredoxin. *Inorg. Chem.* **1994**, *34*, 5358–5369. [CrossRef]

128. Staples, C.R.; Dhawan, I.K.; Finnegan, M.G.; Dwinell, D.A.; Zhou, Z.H.; Huang, H.; Verhagen, M.F.J.M.; Adams, M.W.W.; Johnson, M.K. Electronic, Magnetic, and Redox Properties of [MFe$_3$S$_4$] Clusters (M = Cd, Cu, Cr) in *Pyrococcus furiosus* Ferredoxin. *Inorg. Chem.* **1997**, *36*, 5740–5749. [CrossRef] [PubMed]

129. Ensign, S.A. Reactivity of Carbon Monoxide Dehydrogenase from Rhodospirillum rubrum with Carbon Dioxide, Carbonyl Sulfide, and Carbon Disulfide. *Biochemistry* **1995**, *34*, 5372–5381. [CrossRef] [PubMed]

130. Svetlitchnyi, V.; Peschel, C.; Acker, G.; Meyer, O. Two membrane-associated NiFeS-carbon monoxide dehydrogenases from the anaerobic carbon-monoxide-utilizing eubacterium Carboxydothermus hydrogenoformans. *J. Bacteriol.* **2001**, *183*, 5134–5144. [CrossRef] [PubMed]

131. Dobbek, H.; Gremer, L.; Meyer, O.; Huber, R. Crystal structure and mechanism of CO dehydrogenase, a molybdo iron-sulfur flavoprotein containing S-selanylcysteine. *Proc. Natl. Acad. Sci. USA* **1999**, *96*, 8884–8889. [CrossRef] [PubMed]

132. Meyer, O.; Gremer, L.; Ferner, R.; Ferner, M.; Dobbek, H.; Gnida, M.; Meyer-Klaucke, W.; Huber, R. The role of Se, Mo and Fe in the structure and function of carbon monoxide dehydrogenase. *Biol. Chem.* **2000**, *381*, 865–876. [CrossRef] [PubMed]

133. Ferry, J.G. CO dehydrogenase. *Annu. Rev. Microbiol.* **1995**, *49*, 305–333. [CrossRef]

134. Ermler, U.; Grabarse, W.; Shima, S.; Goubeaud, M.; Thauer, R.K. Active sites of transition-metal enzymes with a focus on nickel. *Curr. Opin. Struct. Biol.* **1998**, *8*, 749–758. [CrossRef]

135. Can, M.; Armstrong, F.A.; Ragsdale, S.W. Structure, Function, and Mechanism of the Nickel Metalloenzymes, CO Dehydrogenase, and Acetyl-CoA Synthase. *Chem. Rev.* **2014**, *114*, 4149–4174. [CrossRef] [PubMed]

136. Jeoung, J.H.; Dobbek, H. Carbon dioxide activation at the Ni,Fe-cluster of anaerobic carbon monoxide dehydrogenase. *Science* **2007**, *318*, 1461–1464. [CrossRef]

137. Busse, S.C.; Mar, G.N.L.; Yu, L.P.; Howard, J.B.; Smith, E.T.; Zhou, Z.H.; Adams, M.W.W. Proton NMR investigation of the oxidized three-iron clusters in the ferredoxins from the hyperthermophilic archae *Pyrococcus furiosus* and *Thermococcus litoralis*. *Biochemistry* **1992**, *31*, 11952–11962. [CrossRef]

138. Calzolai, L.; Zhou, Z.H.; Adams, M.W.W.; Mar, G.N.L. Role of Cluster-Ligated Aspartate in Gating Electron Transfer in the Four-Iron Ferredoxin from the Hyperthermophilic Archaeon Pyrococcus furiosus. *J. Am. Chem. Soc.* **1996**, *118*, 2513–2514. [CrossRef]

139. Telser, J.; Smith, E.T.; Adams, M.W.W.; Conover, R.C.; Johnson, M.K.; Hoffman, B.M. Cyanide Binding to the Novel 4Fe Ferredoxin from Pyrococcus furiosus: Investigation by EPR and ENDOR Spectroscopy. *J. Am. Chem. Soc.* **1995**, *117*, 5133–5140. [CrossRef]

140. Butt, J.N.; Niles, J.; Armstrong, F.A.; Breton, J.; Thomson, A.J. Formation and properties of a stable 'high-potential' copper-iron-sulphur cluster in a ferredoxin. *J. Nat. Struct. Biol.* **1994**, *1*, 427–433. [CrossRef] [PubMed]

141. Srivastava, K.K.P.; Surerus, K.K.; Conover, R.C.; Johnson, M.K.; Park, J.B.; Adams, M.W.W.; Munck, E. Moessbauer study of zinc-iron-sulfur ZnFe$_3$S$_4$ and nickel-iron-sulfur NiFe3S4 clusters in Pyrococcus furiosus ferredoxin. *Inorg. Chem.* **1993**, *32*, 927–936. [CrossRef]

142. Surerus, K.K.; Munck, E.; Moura, I.; Moura, J.J.G.; LeGall, J. Evidence for the formation of a ZnFe$_3$S$_4$ cluster in Desulfovibrio gigas ferredoxin II. *J. Am. Chem. Soc.* **1987**, *109*, 3805–3807. [CrossRef]

143. Martic, M.; Jakab-Simon, I.N.; Haahr, L.T.; Hagen, W.R.; Christensen, H.E. Heterometallic [AgFe$_3$S$_4$] ferredoxin variants: Synthesis, characterization, and the first crystal structure of an engineered heterometallic iron-sulfur protein. *J. Biol. Inorg. Chem.* **2013**, *18*, 261–276. [CrossRef] [PubMed]

144. Fiévet, A.; My, L.; Cascales, E.; Ansaldi, M.; Pauleta, S.R.; Moura, I.; Dermoun, Z.; Bernard, C.S.; Dolla, A.; Aubert, C. The anaerobe-specific orange protein complex of Desulfovibrio vulgaris hildenborough is encoded by two divergent operons coregulated by σ54 and a cognate transcriptional regulator. *J. Bacteriol.* **2011**, *193*, 3207–3219. [CrossRef] [PubMed]

145. Maiti, B.K.; Avilés, T.; Carepo, M.S.P.; Moura, I.; Pauleta, S.R.; Moura, J.J.G. Rearrangement of Mo-Cu-S Cluster Reflects the Structural Instability of Orange Protein Cofactor. *Z. Anorg. Und Allg. Chem. (ZAAC)* **2013**, *639*, 1361–1364. [CrossRef]

146. Najmudin, S.; Bonifácio, C.; Duarte, A.G.; Pauleta, S.R.; Moura, I.; Moura, J.J.; Romão, M.J. Crystallization and crystallographic analysis of the apo form of the orange protein (ORP) from *Desulfovibrio gigas*. *Acta Crystallogr. Sect. F Struct. Biol. Cryst. Commun.* **2009**, *65*, 730–732. [CrossRef]

147. Carepo, M.S.; Pauleta, S.R.; Wedd, A.G.; Moura, J.J.; Moura, I. Mo-Cu metal cluster formation and binding in an orange protein isolated from Desulfovibrio gigas. *J. Biol. Inorg. Chem.* **2014**, *19*, 605–614. [CrossRef]

148. Pauleta, S.R.; Duarte, A.G.; Carepo, M.S.; Pereira, A.S.; Tavares, P.; Moura, I.; Moura, J.J.G. NMR assignment of the apo-form of a Desulfovibrio gigas protein containing a novel Mo-Cu cluster. *Biomol. NMR Assign.* **2007**, *1*, 81–83. [CrossRef] [PubMed]

149. Maiti, B.K.; Moura, J.J.G. Diverse biological roles of the tetrathiomolybdate anion. *Coord. Chem. Rev.* **2021**, *429*, 213635. [CrossRef]

150. Maiti, B.K. A review on chemical and physical properties of tetrathiomolybdate (TTM) anion drug is useful for TTM treated diseases. *J. Indian Chem. Soc.* **2021**, *98*, 100117. [CrossRef]

151. Maiti, B.K.; Pal, K.; Sarkar, S. A structural model of mixed metal sulfide cluster of molybdenum and copper present in the orange protein of Desulfovibrio gigas. *Inorg. Chem. Commun.* **2004**, *7*, 1027–1029. [CrossRef]

152. Maiti, B.K.; Govil, N.; Kundu, T.; Moura, J.J.G. Designed metal-ATCUN derivatives: Redox-and non-redox-based applications relevant for chemistry, biology, and medicine. *Iscience* **2020**, *23*, 101792. [CrossRef]

153. Maiti, B.K.; Avilés, T.; Matzapetakis, M.; Moura, I.; Pauleta, S.R.; Moura, J.J.G. Synthesis of [MoS$_4$]$^{2-}$–M (M = Cu and Cd) Clusters: Potential NMR Spectroscopic Structural Probes for the Orange Protein. *Eur. J. Inorg. Chem.* **2012**, *2012*, 4159–4166. [CrossRef]

154. Maiti, B.K.; Avilés, T.; Moura, I.; Pauleta, S.R.; Moura, J.J.G. Synthesis and characterization of [S$_2$MoS$_2$Cu(n-SPhF)]$^{2-}$ (n = o, m, p) clusters: Potential 19F-NMR structural probes for Orange Protein. *Inorg. Chem. Commun.* **2014**, *45*, 97–100. [CrossRef]

Review

Role of MOB4 in Cell Proliferation and Neurogenesis

Inês B. Santos [1,2,3], Juan Garrido-Maraver [1,2,†], Carolina Gonçalves [1,2,3], Bruna I. Oliveira [1,3] and Álvaro A. Tavares [1,2,3,*]

[1] Faculty of Medicine and Biomedical Sciences, University of Algarve, 8005-139 Faro, Portugal; ibsantos@ualg.pt (I.B.S.); jgarmar1@upo.es (J.G.-M.)
[2] Centre for Biomedical Research, University of Algarve, 8005-139 Faro, Portugal
[3] Algarve Biomedical Center (ABC), University of Algarve, 8005-139 Faro, Portugal
[*] Correspondence: aatavares@ualg.pt
[†] Current address: Centro Andaluz de Biología de Desarollo, Universidad Pablo de Olavide, 41013 Sevilla, Spain.

Abstract: Signaling pathways that integrate a large set of inputs (both extra- and intracellular) to control cell proliferation are essential during both development and adult stages to guarantee organism homeostasis. Mobs are small adaptor proteins that participate in several of these signaling pathways. Here, we review recent advances unravelling Mob4 cellular functions, a highly conserved non-catalytic protein, that plays a diversity of roles in cell proliferation, sperm cell differentiation and is simultaneously involved in synapse formation and neural development. In addition, the gene is often overexpressed in a large diversity of tumors and is linked to poor clinical outcomes. Nevertheless, Mob4 molecular functions remain poorly defined, although it integrates the core structure of STRIPAK, a kinase/phosphatase protein complex, that can act upstream of the Hippo pathway. In this review we focus on the recent findings of Mob4 functions, that have begun to clarify its critical role on cell proliferation and the development of tissues and individuals.

Keywords: cell proliferation; neurogenesis; spermatogenesis; hippo pathway; STRIPAK; Mob4

Citation: Santos, I.B.; Garrido-Maraver, J.; Gonçalves, C.; Oliveira, B.I.; Tavares, Á.A. Role of MOB4 in Cell Proliferation and Neurogenesis. *BioChem* **2023**, *3*, 182–196. https://doi.org/10.3390/biochem3040013

Academic Editor: Buyong Ma

Received: 29 July 2023
Revised: 30 September 2023
Accepted: 27 November 2023
Published: 6 December 2023

1. Introduction

Tissue homeostasis requires a fine balance between cellular proliferation, differentiation and cell death. Throughout evolution, several mechanisms have evolved to ensure the fine-tuning of cell proliferation in multicellular organisms. The Mps-one binder (Mob) family of genes encodes for a highly conserved group of proteins with a central role in regulating some of these mechanisms [1,2]. Mob1p, initially identified in *S. cerevisiae*, was the first Mob gene to be identified and described to be a part of the mitotic exit network (MEN) through its interaction with Dbf2-like kinases [3,4]. In the last twenty five years, Mob homologues have been identified in several model organisms. In eukaryotes, the Mob family is divided into four classes—Mob1, Mob2, Mob3 and Mob4—with non-overlapping functions. Seven Mob genes are encoded in the human genome, all with a high degree of sequence similarity [5,6]. Molecularly, Mob proteins act as adaptor proteins without catalytical activity, that can bind kinases and modulate their activity.

Mob1, the first Mob identified in metazoans, functions as a core component of the Hippo signaling pathway [7], which regulates cell proliferation and organ size [8,9]. Interestingly, the other members of the Mob family can also modulate the activation of the Hippo signaling pathway. For example, Mob2 displays an antagonist function of Mob1 by negatively restricting Hippo signaling [10,11]. On the other hand, in response to apoptotic stimuli and cell–cell contact, Mob3 protects against the induction of apoptosis, thereby sustaining cell proliferation and tumor growth [12]. Finally, Mob4, the focus of this review, can compete with Mob1, thereby restricting Hippo signaling.

The Hippo signaling pathway is a major regulator of cell proliferation, organ size, cellular homeostasis and regeneration. The pathway is evolutionarily conserved and is modulated by a variety of signals such as cell–cell contact, ligands of G-protein coupled receptors, cell polarity, mechanical cues and cellular energy status [13]. The core of the Hippo pathway is composed of a kinase cascade wherein MST1/2 kinases phosphorylates and activates the complex formed by LATS and Mob1, that in turn phosphorylates and inactivates the oncoprotein YAP/TAZ (yes-associated protein 1/tafazzin) that promotes the expression of cell proliferative and antiapoptotic genes [13].

The Hippo pathway is negatively regulated by a large protein complex named STRI-PAK (Striatin-interacting phosphatase and kinase) complex. Core members of STRIPAK include the catalytical (PP2A-C), scaffolding (PP2A-A) and regulatory (Striatins or PP2A-B''') subunits of the serine/threonine protein phosphatase 2A (PP2A), MST3/4 kinases (mammalian sterile 20-like kinases 3 and 4) and the adaptor proteins CCM3 (cerebral cavernous malformation 3), Mob4 and STRIP1/2 (Striatin-interacting proteins 1 and 2) [14–16]. The STRIPAK complex regulates several signaling pathways through the modulation of the phosphorylation levels of its interacting proteins [17,18]. The diversity of proteins associated with STRIPAK highlights its key roles across various biological systems [19,20]. Importantly, Mob4 as a core component of the STRIPAK complex, can restrict Hippo signaling through this second mechanism.

2. MOB4: From the Gene to the Function

Initially named as Phocein [21], it was soon renamed Mob4 because of the high homology it displays with other Mob proteins. The Mob4 gene has been referred in the literature as Mob1, 2C4D, CGI-95, class II mMOB1, Mob1 homolog 3, Mob3, Mps-one binder kinase activator-like 3 (MOBKL3) and preimplantation protein 3 (PREI3). In this work, we follow the HUGO Gene Nomenclature Committee (HGNC) and use Mob4 as the product of the gene ID 17261 (NCBI Entrez Gene: 25843).

Structurally, the canonical Mob fold consists of a four-helix bundle at its core, with three short α-helices at the N-terminal extension (amino acids (aa) 1−61) [22]. Of note, Mob4 share a high structural homology with Mob1, even at the N-terminal, a region where the various Mob proteins diverge.

The human Mob4 gene is predicted to generate alternatively spliced transcriptional variants, producing three predicted protein isoforms: a canonical isoform with 26 kDa (isoform 1) and two smaller variants (Figure 1). One variant results from an alternative exon, containing a different start codon and thus generating a smaller protein with a different N-terminal with 22.3 kDa (isoform 2), and a second variant lacking an in-frame exon generating an isoform with 23.5 kDa (isoform 3) but sharing the N- and C-termini with the canonical isoform. Like all Mob proteins, Mob4 is highly conserved across evolution; for example, there is 80% aa identity and 88% similarity between *Drosophila melanogaster* Mob4 (dMob4) and its human ortholog (hMob4).

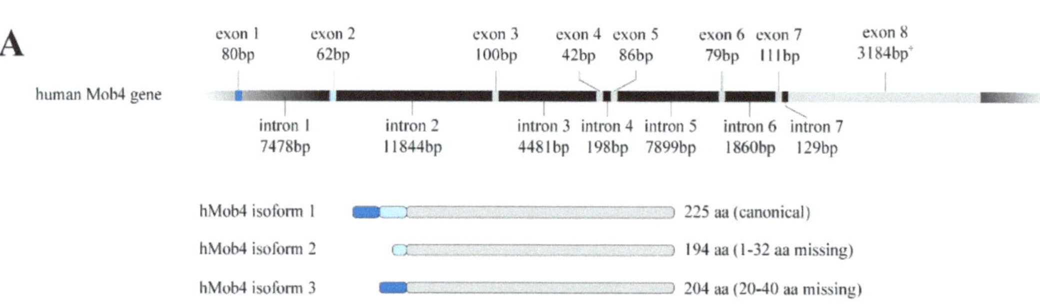

Figure 1. *Cont.*

B

hMob4 isoform 1	1	MVMAEGTAVLRRNRPGTKAQDFYNWPDESFDEMDSTLAVQQYIQQNIRADCSNIDKILEP 60
hMob4 isoform 2	1	------------------------------MDSTLAVQQYIQQNIRADCSNIDKILEP 28
hMob4 isoform 3	1	MVMAEGTAVLRRNRPGTKA--------------------QQYIQQNIRADCSNIDKILEP 39

                                          * * * * * * * * * * * * * * * * * * * *

hMob4 isoform 1	61	PEGQDEGVWKYEHLRQFCLELNGLAVKLQSECHPDTCTQMTATEQWIFLCAAHKTPKECP 120
hMob4 isoform 2	29	PEGQDEGVWKYEHLRQFCLELNGLAVKLQSECHPDTCTQMTATEQWIFLCAAHKTPKECP 88
hMob4 isoform 3	40	PEGQDEGVWKYEHLRQFCLELNGLAVKLQSECHPDTCTQMTATEQWIFLCAAHKTPKECP 99

    *

hMob4 isoform 1	121	AIDYTRHTLDGAACLLNSNKYFPSRVSIKESSVAKLGSVCRRIYRIFSHAYFHHRQIFDE 180
hMob4 isoform 2	89	AIDYTRHTLDGAACLLNSNKYFPSRVSIKESSVAKLGSVCRRIYRIFSHAYFHHRQIFDE 148
hMob4 isoform 3	100	AIDYTRHTLDGAACLLNSNKYFPSRVSIKESSVAKLGSVCRRIYRIFSHAYFHHRQIFDE 159

    *

hMob4 isoform 1	181	YENETFLCHRFTKFVMKYNLMSKDNLIVPILEEEVQNSVSGESEA 225
hMob4 isoform 2	149	YENETFLCHRFTKFVMKYNLMSKDNLIVPILEEEVQNSVSGESEA 193
hMob4 isoform 3	160	YENETFLCHRFTKFVMKYNLMSKDNLIVPILEEEVQNSVSGESEA 204

    *

Figure 1. hMob4 gene structure and predicted encoded proteins. (**A**) hMob4 genetic locus at 2q33.1. hMob4 gene is predicted to generate three different isoforms: the canonical isoform 1 contains 225 aa; Isoform 2 loses the first 32 aa in respect to the canonical isoform 1; Isoform 3 lacks an alternate in-frame exon (exon 2, light blue), generating a smaller isoform (aa 20–40 missing). (**B**) Comparative alignment of the three Mob4 protein isoforms. Dark and light blue regions correspond to exon 1 and 2, respectively. Identical aa are indicated by asterisks.

3. Neuronal Functions of Mob4

Mob4 was initially identified on screenings in neural cell libraries as a potential interactor of Striatins, by the groups of Ariane Monneron and of David Pallas, independently [21,23]. These initial biochemical studies led to the characterization of Mob4 as a core component of the STRIPAK complex (Figure 2) and indicated that Mob4 had a putative role in neuronal function. In agreement with this idea, it was found that mammalian Mob4 is highly expressed in melanized dopamine neurons as well as in the central and peripheral human nervous system [21,24]. In addition, it is highly enriched in dendritic spines, the actin-rich protrusions emerging from dendrites [24–26].

STRIPAK complex

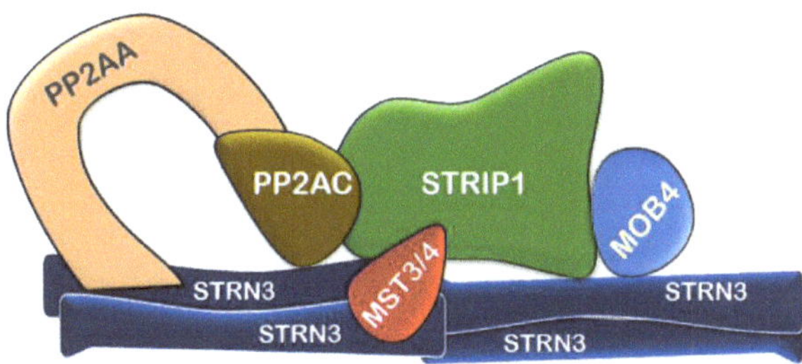

Figure 2. Representative scheme of core mammalian STRIPAK complex. The STRIPAK complex is assembled on a tetramer of Striatin (here represented by STRN3), the scaffolding (PP2A-A) and catalytic (PP2A-C) subunits of phosphatase PP2A, a kinase (MST3 or MST4), and adaptor proteins interacting at different regions of the complex, like Mob4. Mob4 connects STRIP1 to STRN3 (adapted and modified from [22]).

Immunogold labeling detected Mob4 in close vicinity to endocytic-like membranes in the neuronal dendritic spines, suggesting vesicular trafficking functions [24,25]. Immunocytochemical studies also confirmed that Mob4 is strictly somato-dendritic, extending down to the neuronal spines [21]. Noticeably, Mob4 was detected in cell bodies and dendrites but not in axons [21], similarly to what has been described for striatin. In addition, immunoreactivity for Mob4 was only detected in neurons, not in glial cells. Recent studies in zebrafish confirm that Mob4 is highly expressed in the central nervous system; Mob4 was detected in the full extension of the neuroectoderm and in the brain in the initial stages of zebrafish development. Prior to hatching, Mob4 expression is almost limited to the brain, and its expression remains enriched in the brain after hatching [27]. A neural role for Mob4 was first demonstrated by genetic studies in *Drosophila*, where it was shown to be a key regulator of neuronal structure and development [26]. Importantly, these studies showed that Mob4 is an essential gene, as *Drosophila* individuals lacking Mob4 do not survive past the larval stages. The selective somatic downregulation of *Drosophila* Mob4 resulted in a disrupted neuronal morphology. Neurons that developed without Mob4 showed abnormal branching patterns and hyperbranching, and established incorrect connections to target cells. In addition, the lack of Mob4 resulted in severe defects in microtubule organization, synaptic development and axonal transport. Finally, mutant *Drosophila* individuals displayed a defective neuronal transport of cargo from the cell body to the synapses [26]. *dMob4* mutants had neuromere clusters that were smaller in size, although with shorter neurites and a higher thickness of neurite bundles.

Synapses are highly specialized structures that process and transmit information in the brain. These cell–cell communication hubs are thought to be under constant modification during development and by experiences throughout life. Most excitatory synapses localizes to dendritic spines, small actin-rich protrusions on the surface of dendrites. The association of mammalian Mob4 with endocytic-like membranes and neuronal spines suggests a function on endocytosis and vesicular trafficking [24,25]. Intracellular traffic depends on cytoskeleton functioning as rails for the cargo. *dMob4* mutants have a disorganized microtubule cytoskeleton [26], indicating a possible association of Mob4 function with cytoskeleton organization and stability.

Recently, the role of Mob4 in vertebrate neurodevelopment has been highlighted by studying the consequences of Mob4 depletion in zebrafish embryogenesis using morpholinos [27]. The authors found that the knockdown of zfMob4 using translation-blocking morpholinos in young embryos lead to severe neurologic defects. Zebrafish embryos at 24 h post fertilization lacked the midbrain–hindbrain boundary and showed reduced eye size. Mob4 morphants were also shown to be significantly smaller than their control morpholinos siblings. Notably, these differences are not found in the axial trunk but only in the head [27]. This suggests that Mob4 has important functions in the neurodevelopment of the brain but not the spinal cord. In addition, the authors have observed that, in these morphant embryos, not only is the hindbrain and eye regions smaller, but the rate of cell division is severely diminished in the hindbrain and eye regions, arguing that the reduced hindbrain size and eye size are a consequence of impaired cell divisions [27]. Altogether, these results strongly indicate a Mob4 involvement in the process of cell proliferation during neurodevelopment in vertebrates. In agreement with these observations for Mob4 and supporting such functions in neurons, it has been shown that the STRIPAK complex promotes the organization, development and maturation of striatal neurons. In fact, the dysfunction of STRIPAK has been linked to a range of clinical neurological conditions [28]. For example, knockdown of Striatin, that binds directly to Mob4, blocks dendrite formation [29], an observation that suggests that the knockdown of Mob4 affects neuronal morphology at least in part by affecting STRIPAK function.

Taking these observations altogether, Mob4 seems to be required for the regulation of neuronal functions at different levels. Firstly, by being required for the assembly of a normal microtubule cytoskeleton [26]. Second, biochemical studies indicated Mob4 to be involved in endocytosis and vesicular trafficking [24–26]. It seems therefore that Mob4 contributes both to postnatal synaptogenesis and to the dendritic-activity-dependent plasticity in the adult. It is noteworthy to recall that Mob4 is highly conserved throughout the animal kingdom. For this reason, it is tempting to speculate that a deficient Mob4 function may be related to mechanisms of neurological diseases in humans.

4. Mob4 and Cytoskeleton

Recent reports indicate that Mob4 has important roles in cytoskeletal regulation. Studies on *Drosophila* cells reported Mob4 to be involved in mitotic spindle microtubule focusing [30], while studies on the fly nervous system showed Mob4 to be important for the organization of microtubule networks within postmitotic neurons [26]. A recent report shows that in Zebrafish, Mob4 function is required for the incorporation α-actin into organized sarcomeres in skeletal muscle [31]. In this work, the authors found an interaction between Mob4 and the actin-folding chaperonin TRiC (TCP-1 ring complex, also called chaperonin containing TCP-1), suggesting that Mob4 affects TRiC to control actin biogenesis and thus myofibril growth. This report supports a previously described Mob4 molecular interaction with TRiC [14].

The proper folding of proteins is essential for cellular function, and protein misfolding is believed to be the primary cause of many neurodegenerative disorders. In eukaryotes, the folding of misfolded or unfolded proteins is mediated by molecular chaperones, one of which is the protein complex TRiC. Chaperonins are required for the proper folding, transport and degradation of proteins. TRiC chaperonin display limited substrate specificity and assists the folding of many key structural proteins, such as the cytoskeletal proteins actin and α- and β-tubulin [32–34]. The importance of TRiC complex in protein folding was first shown in *Caenorhabditis elegans* where individuals with reduced TRiC function display defective microtubule cytoskeletons [35]. Co-immunoprecipitations assays had shown that Mob4 and TRiC are part of a multiprotein complex and that a direct interaction between Mob4 protein and TRiC was demonstrated to occur in nematodes [14,36]. Recently, Berger et al. showed that Zebrafish Mob4 mutants also display deficient microtubule networks, which is in line with the fact that a main folding substrate of TRiC is tubulin. Therefore, it is possible that such defective microtubule networks are the cause of the observed compromised neuronal connectivity in zebrafish Mob4 mutants. In line with these observations, neuronal neurite formation is strongly affected in TRiC-deficient zebrafish [32,33].

Additionally, Berger et al. also found that in zebrafish, Mob4 protein, co-localized with the marker α-actinin at the sarcomere's Z-discs. Importantly, whereas the loss of Mob4 function led to a smaller amount of myofibrils, increased Mob4 expression induced an increase in the amount of organized myofibrils. These findings indicate that Mob4 function might be required for the regulation of the number of organized myofibrils. Lack of Mob4 function results in reduced numbers of myofibrils and impaired movement due to skeletal muscle defects.

As Mob4 functions within the protein complex STRIPAK, Berger and colleagues looked for similar defects in strn3-deficient mutants. In fact, the authors found that strn3-deficient mutants featured both neuronal and muscle defects. These observations confirm not just Mob4 as a core component of STRIPAK in Zebrafish but also imply a role for the STRIPAK complex in sarcomerogenesis. Unexpectedly, zebrafish TRiC mutants still develop into almost normal larvae that, nonetheless, still show highly specific skeletal muscle defects, resulting from the defective folding of α-actin at Z-disks in the skeletal muscle, with the result of a reduced sarcomere assembly.

The work of Berger et al. highlights the fact that Mob4 function involves at least two different protein complexes, STRIPAK and TRiC. The former has a large diversity of cellular functions, and the latter is required for actin and tubulin biogenesis. *Mob4* and *strn3* mutants featured both neuronal and muscle defects, given that neuronal axons are formed by polar arrays of microtubules, while the major protein components of smooth-muscle thin filaments is actin. This supports the idea that STRIPAK and TRiC may interact through Mob4 to coordinate the growth of the myofibril and of the microtubule network during neural development.

5. Cell Proliferation Function of Mob4

From a physiological point of view, Mob4 plays important roles in the control of cell proliferation. As previously mentioned, the downregulation of zfMob4 results in a severe decrease in the number of mitotic cells in the brain and in the developing eye [27]. The control of stem cell number also seems to be one function of Mob4, as studies on planarian indicated that mob4 function limits body size through limiting stem cell numbers [37]. Planarians have been highly studied for their ability to regenerate body parts after injury and keeping body proportionality [38]. The inhibition of Mob4 in planaria dramatically increased posterior length after injury, affecting the polarity along the apical posterior axis [37].

This an important step forward in deciphering the molecular mechanisms that allow animals to reconstruct body parts while simultaneously integrating newly regenerated tissues into pre-existing old ones after injury. The process of regeneration in planarians is complex and involves the coordination of multiple cellular signaling pathways essential for normal physiological functions, such as organ size, cell proliferation cycle, asymmetric cell divisions, programmed cell death, and cell/tissue polarity determination [39]. One such signaling pathway is the evolutionarily conserved Wnt pathway, known to be involved in a myriad of processes namely cell proliferation, differentiation, and apoptosis, as well as in stem cell maintenance [40]. In planaria, the Wnt signaling pathway is the key regulator of the head–tail polarity, and it is at the core of the decision-making process to regenerate a head or a tail [41,42], as reviewed in [39]. The regionalization of the planarian Antero-Posterior axis is controlled by constitutive localized expression of Wnt ligands (expressed posteriorly) and of Wnt inhibitors (expressed anteriorly). The results obtained by Schad and Petersen support a model in which Mob4 regulates wnt pole cell numbers by limiting stem cell formation, and wnt pole cells in turn control tail proportionality along the antero-posterior axis [37]. These results therefore indicate that Mob4 is involved in the scaling of tail size with respect to body size via the regulation of wnt1. Considering that Mob4 is constitutively expressed, even in the absence of injury, the recovery of normal proportions through regeneration may involve a mechanism that establishes a balance of local signaling processes. These results suggest that the suppression of wnt signaling by Mob4 (through STRIPAK) is a critical pathway regulating scaling and whole-body proportionality in planarians. Interestingly, in planaria, the members of the Hippo pathway do not seem to be required for keeping antero-posterior tissue proportionality, suggesting that Mob4 and STRIPAK exert their action independently of the Hippo signaling pathway (Figure 3). On the other hand, a crosstalk between the Hippo pathway and the Wnt pathway has been previously described [43], and Hippo plays an important role in the regulation of the cell cycle, which is equally crucial for planarian regeneration.

Importantly, Mob proteins have been demonstrated to be involved in defining cell polarity both in humans and in Tetrahymena [44,45]. The importance of cell polarity in a number of physiological processes, including cell differentiation, cell migration, asymmetric cell division, cancer progression and immune response, has been extensively described and reviewed in [46]. In Tetrahymena, a unicellular organism, the single Mob protein encoded in the genome, is required for correct division-plane placement by establishing the anterior–posterior axis. The downregulation of Mob in Tetrahymena induces the misplacement of the division plane with consequent abscission failure; daughter cells fail to separate and form trails of interconnected abnormal cells [44]. Interestingly, the authors found that the Mob protein accumulates at the future site of cell division prior to constriction start, thereby defining the anterior and posterior ends of the future new daughter cells. This finding highlights the importance of Mob in cell polarity inception through a cell-intrinsic mechanism and how polarity is coupled to growth in Tetrahymena. Likewise, in human HeLa cultured cells, the downregulation of Mob1 also results in abscission failure and, importantly, cell polarity is affected in such a way that it allows cells to become motile [45].

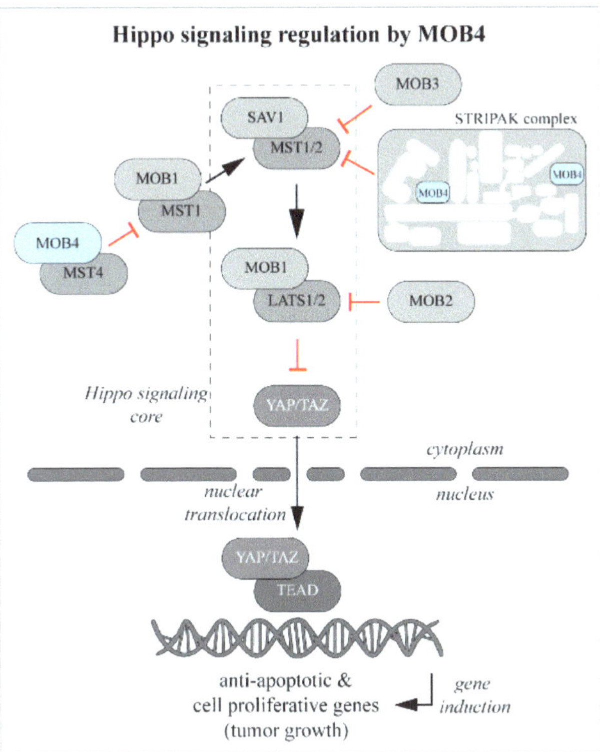

Figure 3. Hippo signaling pathway regulation by Mob proteins. When active, the Hippo pathway blocks cell proliferation. The core of Hippo pathway is a kinase cascade wherein MST1/2, together with SAV1, phosphorylates and activates the complex formed by LATS and Mob1 that, in turn, phosphorylates and inactivates the oncoprotein YAP/TAZ. Active YAP/TAZ (non-phosphorylated) migrates to the nucleus and promotes cell proliferative and antiapoptotic gene. MOB family proteins interact with Hippo pathway at different levels regulating its activity. Mob2 negatively regulates the Hippo pathway by competing with Mob1 for LATS1/2 binding. Mob3 appears to be an MST1 suppressor. Mob1, beside interacting with LATS1/2, can also form a complex with MST1 with tumor-suppressing functions. Mob4 forms a complex with MST4 that antagonizes Mob1-MST1 functions. Mob4 also takes part in the STRIPAK complex, a complex that acts upstream of the Hippo signaling pathway and therefore modulates MST1 activation. The core Hippo pathway is indicated by the traced square. See text for references.

Finally, a role for Mob4 in cell proliferation and tissue formation has also been described in the filamentous fungi *Sodaria macrospora*, where the downregulation of SmMob3 (the Mob4 homologue in *Sodaria*) results in impaired vegetative growth accompanied by a sexually sterile strain unable to undergo self-fusion and fusion [47]. On the other hand, in *Caenorhabditis elegans*, *mob-4*-deficient mutant individuals do not show an obvious abnormal phenotype under the normal growth conditions. Nevertheless, longevity and thermotolerance are affected in these individuals [48]. These observations, together with a previous report showing that *C. elegans* YAP-1 overexpression shortens these individuals' life span (whereas *yap-1* deficiency prolongs life span) [49] suggests that in contrast to humans and *Drosophila*, in which Mob4 activates YAP1 [50], in worms, Mob4 does not act upstream of YAP-1. Taken together, these observations led to suggestion that in *C. elegans*, the Hippo pathway is not conserved [49].

Mob4 may be involved in cell proliferation through its function within the STRIPAK complex or independently of it. STRIPAK negatively regulates the Hippo signaling pathway, thereby participating in the control of cell proliferation [28,51,52]. The deletion of the amino-terminal residues of Mob4 abolishes the assembly of STRIPAK [22] while the disruption of the sites responsible for Mob4 and STRN3 interaction causes aberrant Hippo signaling regulation. Thus, Mob4 can affect cell proliferation by participating in STRIPAK assembly and Hippo pathway activity.

Given the role of Mob4 in regulating cell proliferation it is not surprising that Mob4 may be involved in cancer initiation/progression. Cancer development, due to excessive cell proliferation, is highly associated with the activation of oncogenic pathways [53–55] or the deregulation of genes with tumor- suppressing functions [56–59]. In most human cancers, there is a moderate/high protein expression of Mob4, and a higher expression of Mob4 is associated with a poor prognosis for renal and liver cancers (https://www.proteinatlas.org, accessed on 1 December 2022). The Hippo pathway, per se, is an important regulator of cell proliferation and tissue growth, and mice mutant in Hippo components (Sav (Salvador), MST1/2, Lats (Large tumor suppressor kinase) and Mob1) are prone to develop malignant growths [60–63]. In addition, it has been described that the different Mob family members behave either as tumor suppressors or oncogenes. For example, the complete loss of Mob1, a component of the Hippo pathway, in mice promotes tumorigenesis and embryonic death [60]; Mob2 has recently been reported as a tumor suppressor in glioblastoma [64]; and in contrast, Mob3 has been found to be upregulated in glioblastoma multiforme, and it is proposed as an oncoprotein by suppressing MST1 activity [12].

Interestingly, Mob4 can also bind to the protein kinase MST4 forming a complex that antagonizes the complex formed by Mob1 and the kinases MST1/2 [50] (Figure 3). But while MST1-Mob1 acts as a tumor suppressor, MST4-Mob4 is oncogenic by activating YAP signaling and promoting cell proliferation. In fact, Mob4 can alternatively pair with either MST4 or MST1 due to the high structural similarities of both Mobs and of both MST kinases. Mob4 can therefore sequester MST1, consequently inhibiting the Hippo pathway and promoting cell proliferation [50], thus acting as an oncogene. It is worth remembering that simultaneously, on an alternative mechanism and as a component of STRIPAK complex, Mob4 negatively regulates the Hippo signaling pathway and therefore acts as a tumor suppressor gene.

6. Mob4 and Spermatogenesis

Spermatogenesis is the process by which germinal stem cells give rise to haploid spermatozoa. In *Drosophila*, spermatids morphogenesis generally occurs within a syncytium, with all spermatid nuclei remaining interconnected via an extensive network of cytoplasmic bridges. As spermiogenesis progresses, the syncytium (or cyst) is then resolved into individual cells in a process referred to as sperm individualization. Although with some differences in the control of hormonal regulation and in testicular structure, the different stages of spermatogenesis are highly conserved from fly to human. Importantly, many of the genes involved in *Drosophila* spermatogenesis were shown to be conserved in humans.

Santos et al. recently described a requirement for Mob4 during spermatogenesis in *Drosophila*: males without Mob4 function in the gonads are sterile, while females are fully fertile. In the lack of Mob4 function, cyst elongation still occurs, meaning that spermatids are capable of elongating an axoneme, but spermatid individualization was observed to fail. Consequently, the migration of sperm into the seminal vesicle does not occur, and consequently, males are sterile. In order to determine what defects in spermiogenesis cause spermatid individualization failure, the authors have examined the ultrastructure of the developing axoneme and found that the mitochondrial derivatives fail to form correctly. Importantly, defects in the axonemal structure were also found. These defects include the loss of microtubule doublets, and most interestingly, the stereotypical 9 + 2 microtubule doublets of the axoneme is affected, with the axoneme suffering a large radial expansion [65].

The use of a GFP:Mob4 transgene reveals that Mob4 has dynamic sub-cellular localization in different cell types throughout spermatogenesis. During meiosis, Mob4 accumulates in the reticulum surrounding the meiotic spindle structures. Post-meiotically, Mob4 transiently accumulates at the basal side of the nuclei in the vicinity of the basal body at the early canoe stage of spermatid differentiation. The dynamic behavior of Mob4 throughout different stages of spermatogenesis is suggestive of multiple roles in the parafusorial membranes and associated microtubules during meiosis, and at the basal body or transition zone in the initiation of axoneme elongation.

As mentioned, spermatogenesis is a highly conserved process, and the genes involved in spermatogenesis usually have their function conserved across species. Mob4 seem to follow the trend as the insertion of the human Mob4 paralog gene into *Drosophila* was capable of rescuing all the meiotic defects in mob4 mutant, including full fertility [65], suggesting that human Mob4 and *Drosophila* Mob4 are functional orthologs.

Considering that Mob4 is a component of STRIPAK, one may wonder if the defective spermatogenesis results from lack of STRIPAK function, or is an isolated function of Mob4. To answer this question Santos et al. looked for a function of Strip and Cka (two other components of STRIPAK) in *Drosophila* testes. They found that, like for Mob4, either Strip or Cka downregulation in testes results equally in male sterility; on the other hand, Strip or Cka downregulation did not seem to affect female ovaries and are not required for female oogenesis. In addition, the investigators also showed that similar failures in sperm individualization to Mob4 are observed after the downregulation of either Strip or Cka, suggesting that STRIPAK complex activity is required for spermatogenesis, and that Mob4 is probably acting through STRIPAK.

7. Other Mob4 Functions

A different set of studies indicated that Mob4 is also required for mitosis progression. In *Drosophila* S2 cultured cells, Mob4 was found to associate with centrosomes and kinetochores, and its downregulation resulted in the formation of monopolar spindles and defective mitosis [30]. Moreover, in human cells, Frost and collaborators showed that the downregulation of Mob4 causes an increase in DNA content, abnormal spindle formation and mitosis failure, triggering cell death [66]. In addition, the authors found that Mob4, together with the STRIPAK complex, bridges the centrosome with the cis-Golgi and the outer nuclear membrane [21,66] (Figure 4). The disruption of these bridges may be the cause of the defective mitotic progression.

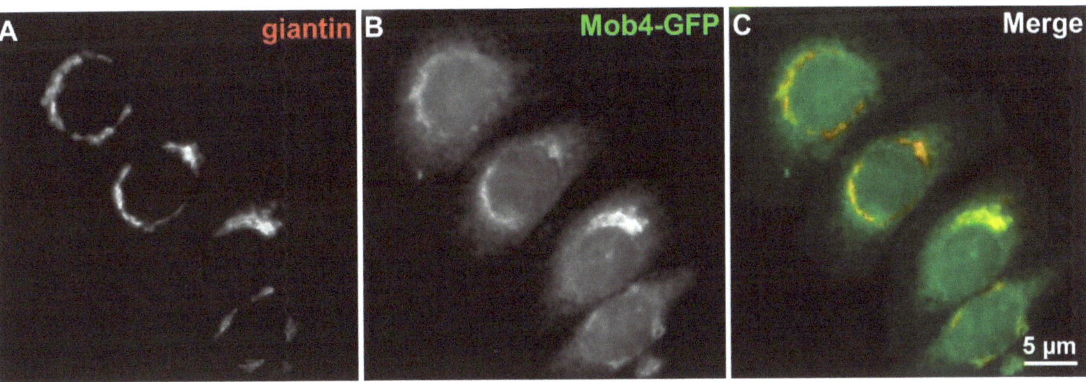

Figure 4. Mob4 accumulates on the Golgi apparatus in human cells [21]. HeLa cells endogenously expressing GFP-tagged Mob4 (**B**) were immunostained with anti-giantin (**A**) (BioLegend, Poly19087) to reveal the Golgi complex. Left and middle panel are single channel images and right panel (**C**) is the merged image (photos by Inês Santos and Álvaro Tavares).

Human Mob4 may also have a role in DNA damage signaling since its downregulation results in γH2AX phosphorylation in cells [67], which is an early marker for DNA damage [68]. Moreover, the Ser147 of human Mob4 is a possible target of ATM kinase suggesting a possible role for Mob4 within the DNA damage response [69]. Other MOB family members are also involved in the DNA damage signaling. For example, Mob2 can interact with RAD50 promoting the assembly of the MRN DNA damage sensor complex this way activating ATM kinase, a well-known protein orchestrating DNA damage response. In addition, Mob2 competes with Mob1 for NDR1/2 binding (NDR–Nuclear Dbf2-Related Kinase), and NDR-mediated phosphorylation is important in the G2/M DNA damage checkpoint by promoting the degradation of the CDC25A phosphatase [70].

Apoptosis is a process important for the maintenance of cell numbers, and its regulation has been previously associated with Mob4 and Mob1 due to their association with MST1 kinase. MST1 modulates oxidative-stress-induced neuronal death [71], and several studies have reported how the MST1 phosphorylation of FOXO proteins enhances their nuclear translocation, promoting the transcription of apoptosis-related genes [72,73]. A neuronal-specific isoform of YAP (the end target of the Hippo pathway), YAPdeltaC, acts as a neuronal apoptosis protector that decreases with progression for amyotrophic lateral sclerosis (ALS), whereas the full-length YAP remains constant during the late and severe stage of the disease [74].

Specific connections between Mob proteins and neurodegeneration can be established through their interaction with NDR/LATS kinases. It was found that all four *Drosophila* Mob genes can genetically interact with tricornered (Trc) (*Drosophila* homologue of NDR kinases) [75]. Trc is required for morphological changes, such as the outgrowth of epidermal hair and dendritic tiling in sensory neurons and Wts (*Drosophila* homologue of LATS) plays a role in dendritic maintenance in sensory neurons [76–78]. Mob2 has been directly associated with neuronal functions. In neuronal cell lines, Mob2 is required for sustaining neurite formation [79]. Mob2 expression is required to regulate the growth of *Drosophila* larval neuromuscular junction [80] and for normal neuronal distribution in mice developmental cortex [81]. In addition, human Mob2 has been identified as a specific protein of cerebral amyloid angiopathy (CAA), a condition in which amyloid plaques are deposited on the walls of cortical and blood vessels of the brain [82]. On the other hand, increasing evidence shows a potential role for Mob3 in neurodegenerative diseases. The human Mob3A gene was identified as a target for the nuclear respiratory factor 1 (NRF1) [83]. NRF1 is a master transcription factor that promotes the transcriptional activation of genes required for mitochondrial biogenesis and proteosome function. The deregulation of NRF1 target genes has been shown in neurodegenerative disease models where proteosome capacity is diminished [84]. On a second study, human Mob3 proteins were found to bind to the oligomeric Aβ_{42}, a hallmark protein complex of Alzeimer's disease, suggesting the potential involvement of Mob3 proteins with this condition [85]. Therefore, Mob4 is not the only Mob with a role in the control of cell proliferation and apoptosis, which is relevant for the clearance of excessive neural cells during normal development [86]. Finally, very recently, Guo et al. [87] suggested a new function for Mob4, through its role within STRIPAK, in the regulation of autophagy in *Drosophila* muscle tissue.

In summary, Mob family proteins have important functions in the control of cell proliferation. Mob4 displays a range of essential functions that spans from neuronal development to spermatogenesis and the control of cell proliferation, both in vertebrates and in *Drosophila* (Figure 5). These functions are likely exerted through different mechanisms. Mob4 can act as a scaffold on the assembly of the STRIPAK complex (Figures 2 and 3), thus participating in axonal transport, dendritic development and synapse assembly. But Mob4 also appears to act independently of STRIPAK by interacting with the TRiC complex or directly with MST1. Most human cancers show a moderate/high protein expression of Mob4, and renal and liver cancer patients with a high expression of Mob4 show a reduced survival probability compared to those with low expression, suggesting that Mob4 is an unfavorable prognostic marker for renal and liver cancers. Therefore, elucidating the underlying mechanisms of Mob4 action and regulation may be of help to identify novel therapeutic targets and diagnostic markers for cancer and neurodegenerative diseases.

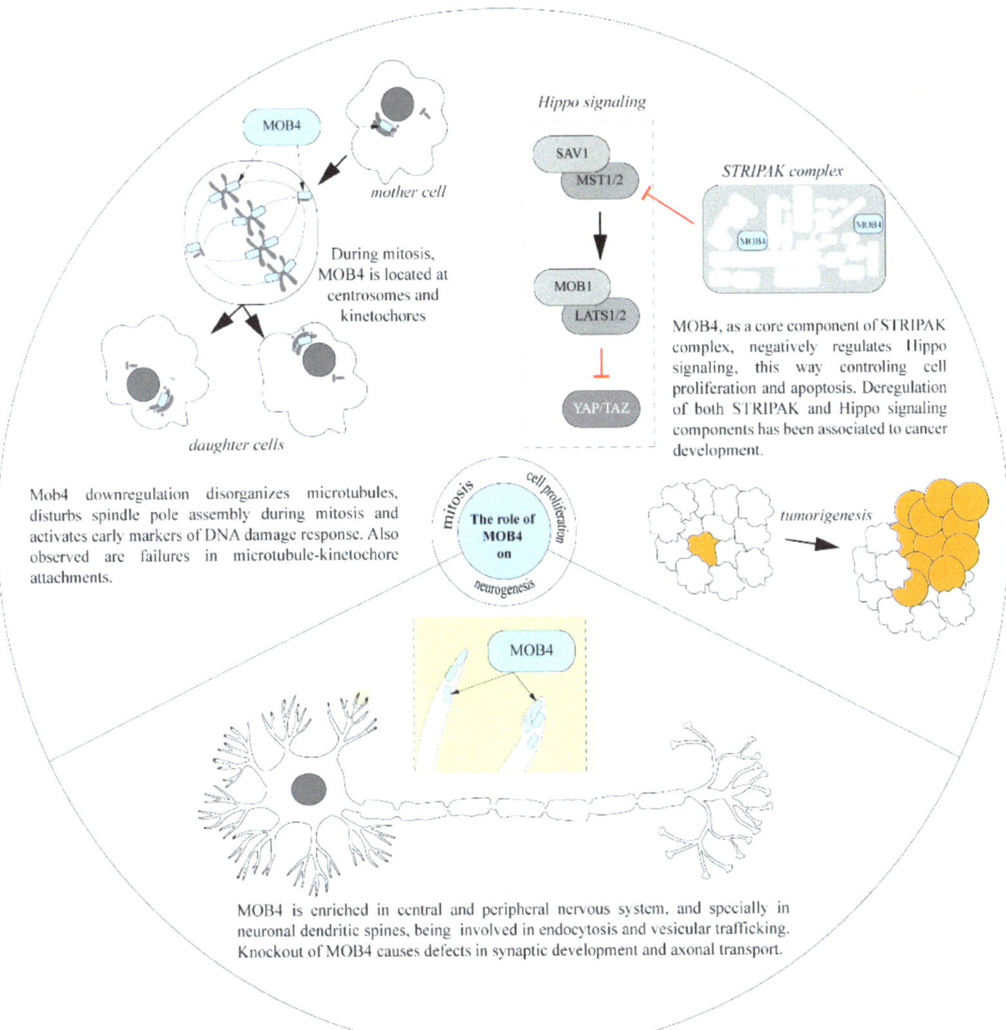

Figure 5. An integrated view of Mob4 functions (mitosis, cell proliferation and neurogenesis). Mob4 has been reported to participate in mitotic spindle assembly and cell division; in the control of cell proliferation regulating the Hippo signaling pathway as a member of STRIPAK complex; and in neural development, regulating axonal transport, dendrite branching and controlling apoptosis in neurons.

Author Contributions: Original draft preparation, writing—review and editing: I.B.S., J.G.-M., C.G., B.I.O. and Á.A.T.; funding acquisition: Á.A.T. All authors have read and agreed to the published version of the manuscript.

Funding: Work funded by the Algarve 2020 Program, grant number ALG-01-0145-FEDER-030014, and co-financed by FEDER Funds through the Operational Program for Competitiveness Factors—COMPETE 2020 and by National Funds through FCT—Foundation for Science and Technology under the Project PTDC/BIA-CEL/30014/2017. I.B.Santos was funded by FCT fellowship SFRH/BD/141734/2018.

Institutional Review Board Statement: Not applicable.

Informed Consent Statement: Not applicable.

Data Availability Statement: Data are contained within the article and references.

Conflicts of Interest: The authors declare no conflict of interest.

References

1. Delgado, I.; Carmona, B.; Nolasco, S.; Santos, D.; Leitão, A.; Soares, H. MOB: Pivotal Conserved Proteins in Cytokinesis, Cell Architecture and Tissue Homeostasis. *Biology* **2020**, *9*, 413. [CrossRef] [PubMed]
2. Duhart, J.C.; Raftery, L.A. Mob Family Proteins: Regulatory Partners in Hippo and Hippo-Like Intracellular Signaling Pathways. *Front. Cell Dev. Biol.* **2020**, *8*, 161. [CrossRef] [PubMed]
3. Luca, F.C.; Winey, M. Regulation of Mob1p, an essential budding yeast protein required for completion of mitosis and spindle pole body duplication. *Mol. Biol. Cell* **1998**, *9*, 12A. [CrossRef] [PubMed]
4. Luca, F.C.; Winey, M. *MOB1*, an essential yeast gene required for completion of mitosis and maintenance of ploidy. *Mol. Biol. Cell* **1998**, *9*, 29–46. [CrossRef]
5. Ye, X.; Nikolaidis, N.; Nei, M.; Lai, Z. Evolution of the mob Gene Family. *Open Cell Signal. J.* **2009**, *1*, 1–11. [CrossRef]
6. Vitulo, N.; Vezzi, A.; Galla, G.; Citterio, S.; Marino, G.; Ruperti, B.; Zermiani, M.; Albertini, E.; Valle, G.; Barcaccia, G. Characterization and evolution of the cell cycle-associated mob domain-containing proteins in eukaryotes. *Evol. Bioinform.* **2007**, *3*, 121–158. [CrossRef]
7. Lai, Z.C.; Wei, X.; Shimizu, T.; Ramos, E.; Rohrbaugh, M.; Nikolaidis, N.; Ho, L.L.; Li, Y. Control of cell proliferation and apoptosis by mob as tumor suppressor, mats. *Cell* **2005**, *120*, 675–685. [CrossRef]
8. Pan, D. Hippo signaling in organ size control. *Genes Dev.* **2007**, *21*, 886–897. [CrossRef]
9. Halder, G.; Johnson, R.L. Hippo signaling: Growth control and beyond. *Development* **2011**, *138*, 9–22. [CrossRef]
10. Weiss, E.L.; Kurischko, C.; Zhang, C.; Shokat, K.; Drubin, D.G.; Luca, F.C. The *Saccharomyces cerevisiae* Mob2p-Cbk1p kinase complex promotes polarized growth and acts with the mitotic exit network to facilitate daughter cell-specific localization of Ace2p transcription factor. *J. Cell Biol.* **2002**, *158*, 885–900. [CrossRef]
11. Kohler, R.S.; Schmitz, D.; Cornils, H.; Hemmings, B.A.; Hergovich, A. Differential NDR/LATS interactions with the human MOB family reveal a negative role for human MOB2 in the regulation of human NDR kinases. *Mol. Cell. Biol.* **2010**, *30*, 4507–4520. [CrossRef]
12. Tang, F.; Zhang, L.; Xue, G.; Hynx, D.; Wang, Y.; Cron, P.D.; Hundsrucker, C.; Hergovich, A.; Frank, S.; Hemmings, B.A.; et al. hMOB3 modulates MST1 apoptotic signaling and supports tumor growth in glioblastoma multiforme. *Cancer Res* **2014**, *74*, 3779–3789. [CrossRef]
13. Ma, S.; Meng, Z.; Chen, R.; Guan, K.L. The Hippo Pathway: Biology and Pathophysiology. *Annu. Rev. Biochem.* **2019**, *88*, 577–604. [CrossRef] [PubMed]
14. Goudreault, M.; D'Ambrosio, L.M.; Kean, M.J.; Mullin, M.J.; Larsen, B.G.; Sanchez, A.; Chaudhry, S.; Chen, G.I.; Sicheri, F.; Nesvizhskii, A.I.; et al. A PP2A phosphatase high density interaction network identifies a novel striatin-interacting phosphatase and kinase complex linked to the cerebral cavernous malformation 3 (CCM3) protein. *Mol. Cell. Proteom.* **2009**, *8*, 157–171. [CrossRef] [PubMed]
15. Glatter, T.; Wepf, A.; Aebersold, R.; Gstaiger, M. An integrated workflow for charting the human interaction proteome: Insights into the PP2A system. *Mol. Syst. Biol.* **2009**, *5*, 237. [CrossRef] [PubMed]
16. Ribeiro, P.S.; Josué, F.; Wepf, A.; Wehr, M.C.; Rinner, O.; Kelly, G.; Tapon, N.; Gstaiger, M. Combined functional genomic and proteomic approaches identify a PP2A complex as a negative regulator of Hippo signaling. *Mol. Cell* **2010**, *39*, 521–534. [CrossRef] [PubMed]
17. Gordon, J.; Hwang, J.; Carrier, K.J.; Jones, C.A.; Kern, Q.L.; Moreno, C.S.; Karas, R.H.; Pallas, D.C. Protein phosphatase 2a (PP2A) binds within the oligomerization domain of striatin and regulates the phosphorylation and activation of the mammalian Ste20-Like kinase Mst3. *BMC Biochem.* **2011**, *12*, 54. [CrossRef]
18. Hornbeck, P.V.; Zhang, B.; Murray, B.; Kornhauser, J.M.; Latham, V.; Skrzypek, E. PhosphoSitePlus, 2014: Mutations, PTMs and recalibrations. *Nucleic Acids Res.* **2015**, *43*, D512–D520. [CrossRef]
19. Gil-Ranedo, J.; Gonzaga, E.; Jaworek, K.J.; Berger, C.; Bossing, T.; Barros, C.S. STRIPAK Members Orchestrate Hippo and Insulin Receptor Signaling to Promote Neural Stem Cell Reactivation. *Cell Rep.* **2019**, *27*, 2921–2933.e5. [CrossRef]
20. Shi, Z.; Jiao, S.; Zhou, Z. STRIPAK complexes in cell signaling and cancer. *Oncogene* **2016**, *35*, 4549–4557. [CrossRef]
21. Baillat, G.; Moqrich, A.; Castets, F.; Baude, A.; Bailly, Y.; Benmerah, A.; Monneron, A. Molecular cloning and characterization of phocein, a protein found from the Golgi complex to dendritic spines. *Mol. Biol. Cell* **2001**, *12*, 663–673. [CrossRef]

22. Jeong, B.C.; Bae, S.J.; Ni, L.; Zhang, X.; Bai, X.C.; Luo, X. Cryo-EM structure of the Hippo signaling integrator human STRIPAK. *Nat. Struct. Mol. Biol.* **2021**, *28*, 290–299. [CrossRef]

23. Moreno, C.S.; Lane, W.S.; Pallas, D.C. A mammalian homolog of yeast MOB1 is both a member and a putative substrate of striatin family-protein phosphatase 2A complexes. *J. Biol. Chem.* **2001**, *276*, 24253–24260. [CrossRef]

24. Haeberlé, A.M.; Castets, F.; Bombarde, G.; Baillat, G.; Bailly, Y. Immunogold localization of MOB4 in dendritic spines. *J. Comp. Neurol.* **2006**, *495*, 336–350. [CrossRef] [PubMed]

25. Bailly, Y.J.R.; Castets, F. Phocein: A potential actor in vesicular trafficking at Purkinje cell dendritic spines. *Cerebellum* **2007**, *6*, 344–352. [CrossRef]

26. Schulte, J.; Sepp, K.J.; Jorquera, R.A.; Wu, C.; Song, Y.; Hong, P.; Littleton, J.T. DMob4/Phocein regulates synapse formation, axonal transport, and microtubule organization. *J. Neurosci.* **2010**, *30*, 5189–5203. [CrossRef] [PubMed]

27. Florindo, C.; Mimoso, J.M.; Palma, S.L.; Gonçalves, C.; Silvestre, D.; Campinho, M.; Tavares, Á.A. Mob4 is required for neurodevelopment in zebrafish. *microPubl. Biol.* **2023**. [CrossRef]

28. Hwang, J.; Pallas, D.C. STRIPAK complexes: Structure, biological function, and involvement in human diseases. *Int. J. Biochem. Cell Biol.* **2014**, *47*, 118–148. [CrossRef]

29. Li, D.; Musante, V.; Zhou, W.; Picciotto, M.R.; Nairn, A.C. Striatin-1 is a B subunit of protein phosphatase PP2A that regulates dendritic arborization and spine development in striatal neurons. *J. Biol. Chem.* **2018**, *293*, 11179–11194. [CrossRef]

30. Trammell, M.A.; Mahoney, N.M.; Agard, D.A.; Vale, R.D. Mob4 plays a role in spindle focusing in *Drosophila* S2 cells. *J. Cell Sci.* **2008**, *121*, 1284–1292. [CrossRef]

31. Berger, J.; Berger, S.; Currie, P.D. Mob4-dependent STRIPAK involves the chaperonin TRiC to coordinate myofibril and microtubule network growth. *PLoS Genet.* **2022**, *18*, e1010287. [CrossRef] [PubMed]

32. Yaffe, M.B.; Farr, G.W.; Miklos, D.; Horwich, A.L.; Sternlicht, M.L.; Sternlicht, H. TCP1 complex is a molecular chaperone in tubulin biogenesis. *Nature* **1992**, *358*, 245–248. [CrossRef] [PubMed]

33. Berger, J.; Berger, S.; Li, M.; Jacoby, A.S.; Arner, A.; Bavi, N.; Stewart, A.G.; Currie, P.D. In Vivo Function of the Chaperonin TRiC in α-Actin Folding during Sarcomere Assembly. *Cell Rep.* **2018**, *22*, 313–322. [CrossRef] [PubMed]

34. Ghozlan, H.; Cox, A.; Nierenberg, D.; King, S.; Khaled, A.R. The TRiCky Business of Protein Folding in Health and Disease. *Front. Cell Dev. Biol.* **2022**, *10*, 906530. [CrossRef] [PubMed]

35. Saegusa, K.; Sato, M.; Sato, K.; Nakajima-Shimada, J.; Harada, A.; Sato, K.; Geisler, F.; Gerhardus, H.; Carberry, K.; Davis, W.; et al. *Caenorhabditis elegans* chaperonin CCT/TRiC is required for actin and tubulin biogenesis and microvillus formation in intestinal epithelial cells. *Mol. Biol. Cell* **2014**, *25*, 3095–3104. [CrossRef]

36. Khabirova, E.; Moloney, A.; Marciniak, S.J.; Williams, J.; Lomas, D.A.; Oliver, S.G.; Favrin, G.; Sattelle, D.B.; Crowther, D.C. The TRiC/CCT chaperone is implicated in Alzheimer's disease based on patient GWAS and an RNAi screen in Abeta-expressing *Caenorhabditis elegans. PLoS ONE* **2014**, *9*, e102985. [CrossRef] [PubMed]

37. Schad, E.G.; Petersen, C.P. STRIPAK Limits Stem Cell Differentiation of a WNT Signaling Center to Control Planarian Axis Scaling. *Curr. Biol.* **2020**, *30*, 254–263.e2. [CrossRef]

38. Wagner, D.E.; Wang, I.E.; Reddien, P.W. Clonogenic neoblasts are pluripotent adult stem cells that underlie planarian regeneration. *Science* **2011**, *332*, 811–816. [CrossRef]

39. Reddien, P.W. The Cellular and Molecular Basis for Planarian Regeneration. *Cell* **2018**, *175*, 327–345. [CrossRef]

40. Clevers, H.; Loh, K.M.; Nusse, R. Stem cell signaling. An integral program for tissue renewal and regeneration: Wnt signaling and stem cell control. *Science* **2014**, *346*, 1248012. [CrossRef]

41. Gurley, K.A.; Rink, J.C.; Sanchez Alvarado, A. Beta-catenin defines head versus tail identity during planarian regeneration and homeostasis. *Science* **2008**, *319*, 323–327. [CrossRef] [PubMed]

42. Petersen, C.P.; Reddien, P.W. Smed-betacatenin-1 is required for anteroposterior blastema polarity in planarian regeneration. *Science* **2008**, *319*, 327–330. [CrossRef] [PubMed]

43. Sileo, P.; Simonin, C.; Melnyk, P.; Chartier-Harlin, M.C.; Cotelle, P. Crosstalk between the Hippo Pathway and the Wnt Pathway in Huntington's Disease and Other Neurodegenerative Disorders. *Cells* **2022**, *11*, 3631. [CrossRef] [PubMed]

44. Tavares, A.; Gonçalves, J.; Florindo, C.; Tavares, A.A.; Soares, H. Mob1: Defining cell polarity for proper cell division. *J. Cell Sci.* **2012**, *125*, 516–527. [CrossRef] [PubMed]

45. Florindo, C.; Perdigão, J.; Fesquet, D.; Schiebel, E.; Pines, J.; Tavares, A.A. Human Mob1 proteins are required for cytokinesis by controlling microtubule stability. *J. Cell Sci.* **2012**, *125*, 3085–3090. [CrossRef] [PubMed]

46. Piroli, M.E.; Blanchette, J.O.; Jabbarzadeh, E. Polarity as a physiological modulator of cell function. *Front. Biosci.* **2019**, *24*, 451–462. [CrossRef]

47. Bernhards, Y.; Pöggeler, S. The phocein homologue SmMOB3 is essential for vegetative cell fusion and sexual development in the filamentous ascomycete *Sordaria macrospora. Curr. Genet.* **2011**, *57*, 133–149. [CrossRef] [PubMed]

48. Jahan, M.; Iwasa, H.; Kuroyanagi, H.; Hata, Y. Loss of *Caenorhabditis elegans* homologue of human MOB4 compromises life span, health life span and thermotolerance. *Genes Cells* **2021**, *26*, 798–806. [CrossRef]

49. Iwasa, H.; Maimaiti, S.; Kuroyanagi, H.; Kawano, S.; Inami, K.; Timalsina, S.; Ikeda, M.; Nakagawa, K.; Hata, Y. Yes-associated protein homolog, YAP-1, is involved in the thermotolerance and aging in the nematode *Caenorhabditis elegans. Exp. Cell Res.* **2013**, *319*, 931–945. [CrossRef]

50. Chen, M.; Zhang, H.; Shi, Z.; Li, Y.; Zhang, X.; Gao, Z.; Zhou, L.; Ma, J.; Xu, Q.; Guan, J.; et al. The MST4-MOB4 complex disrupts the MST1–MOB1 complex in the Hippo–YAP pathway and plays a pro-oncogenic role in pancreatic cancer. *J. Biol. Chem.* **2018**, *293*, 14455–14469. [CrossRef]

51. Pan, D. The hippo signaling pathway in development and cancer. *Dev. Cell* **2010**, *19*, 491–505. [CrossRef] [PubMed]

52. Kück, U.; Radchenko, D.; Teichert, I. STRIPAK, a highly conserved signaling complex, controls multiple eukaryotic cellular and developmental processes and is linked with human diseases. *Biol. Chem.* **2019**, *400*, 1005–1022. [CrossRef]

53. Chiu, I.M. Growth factor genes as oncogenes. *Mol. Chem. Neuropathol.* **1989**, *10*, 37–52. [CrossRef]

54. Miller, M.S.; Miller, L.D. RAS Mutations and Oncogenesis: Not all RAS Mutations are Created Equally. *Front. Genet.* **2012**, *2*, 100. [CrossRef] [PubMed]

55. Morrish, F.; Neretti, N.; Sedivy, J.M.; Hockenbery, D.M. The oncogene c-Myc coordinates regulation of metabolic networks to enable rapid cell cycle entry. *Cell Cycle* **2008**, *7*, 1054–1066. [CrossRef] [PubMed]

56. Goodrich, D.W. The retinoblastoma tumor-suppressor gene, the exception that proves the rule. *Oncogene* **2006**, *25*, 5233–5243. [CrossRef]

57. Smith, A.L.; Robin, T.P.; Ford, H.L. Molecular pathways: Targeting the TGF-β pathway for cancer therapy. *Clin. Cancer Res.* **2012**, *18*, 4514–4521. [CrossRef]

58. Savage, K.I.; Harkin, D.P. BRCA1, a 'complex' protein involved in the maintenance of genomic stability. *FEBS J.* **2015**, *282*, 630–646. [CrossRef]

59. Mantovani, F.; Collavin, L.; Del Sal, G. Mutant p53 as a guardian of the cancer cell. *Cell Death Differ.* **2019**, *26*, 199–212. [CrossRef]

60. Nishio, M.; Hamada, K.; Kawahara, K.; Sasaki, M.; Noguchi, F.; Chiba, S.; Mizuno, K.; Suzuki, S.O.; Dong, Y.; Tokuda, M.; et al. Cancer susceptibility and embryonic lethality in Mob1a/1b double-mutant mice. *J. Clin. Investig.* **2012**, *122*, 4505–4518. [CrossRef]

61. St John, M.A.; Tao, W.; Fei, X.; Fukumoto, R.; Carcangiu, M.L.; Brownstein, D.G.; Parlow, A.F.; McGrath, J.; Xu, T. Mice deficient of Lats1 develop soft-tissue sarcomas, ovarian tumours and pituitary dysfunction. *Nat. Genet.* **1999**, *21*, 182–186. [CrossRef] [PubMed]

62. Zhou, D.; Conrad, C.; Xia, F.; Park, J.S.; Payer, B.; Yin, Y.; Lauwers, G.Y.; Thasler, W.; Lee, J.T.; Avruch, J.; et al. Mst1 and Mst2 maintain hepatocyte quiescence and suppress hepatocellular carcinoma development through inactivation of the Yap1 oncogene. *Cancer Cell* **2009**, *16*, 425–438. [CrossRef]

63. Lee, K.P.; Lee, J.H.; Kim, T.S.; Kim, T.H.; Park, H.D.; Byun, J.S.; Kim, M.C.; Jeong, W.I.; Calvisi, D.F.; Kim, J.M.; et al. The Hippo-Salvador pathway restrains hepatic oval cell proliferation, liver size, and liver tumorigenesis. *Proc. Natl. Acad. Sci. USA* **2010**, *107*, 8248–8253. [CrossRef] [PubMed]

64. Jiang, K.; Yao, G.; Hu, L.; Yan, Y.; Liu, J.; Shi, J.; Chang, Y.; Zhang, Y.; Liang, D.; Shen, D.; et al. MOB2 suppresses GBM cell migration and invasion via regulation of FAK/Akt and cAMP/PKA signaling. *Cell Death Dis.* **2020**, *11*, 230. [CrossRef]

65. Santos, I.B.; Wainman, A.; Garrido-Maraver, J.; Pires, V.; Riparbelli, M.G.; Kovács, L.; Callaini, G.; Glover, D.M.; Tavares, Á.A. Mob4 is essential for spermatogenesis in *Drosophila melanogaster*. *Genetics* **2023**, *224*, iyad104. [CrossRef]

66. Frost, A.; Elgort, M.G.; Brandman, O.; Ives, C.; Collins, S.R.; Miller-Vedam, L.; Weibezahn, J.; Hein, M.Y.; Poser, I.; Mann, M.; et al. Functional repurposing revealed by comparing *S. pombe* and *S. cerevisiae* genetic interactions. *Cell* **2012**, *149*, 1339–1352. [CrossRef] [PubMed]

67. Paulsen, R.D.; Soni, D.V.; Wollman, R.; Hahn, A.T.; Yee, M.C.; Guan, A.; Hesley, J.A.; Miller, S.C.; Cromwell, E.F.; Solow-Cordero, D.E.; et al. A genome-wide siRNA screen reveals diverse cellular processes and pathways that mediate genome stability. *Mol. Cell* **2009**, *35*, 228–239. [CrossRef] [PubMed]

68. Mah, L.J.; El-Osta, A.; Karagiannis, T.C. γH2AX: A sensitive molecular marker of DNA damage and repair. *Leukemia* **2010**, *24*, 679–686. [CrossRef]

69. Wong, Y.H.; Lee, T.Y.; Liang, H.K.; Huang, C.M.; Wang, T.Y.; Yang, Y.H.; Chu, C.H.; Huang, H.D.; Ko, M.T.; Hwang, J.K. KinasePhos 2.0: A web server for identifying protein kinase-specific phosphorylation sites based on sequences and coupling patterns. *Nucleic Acids Res.* **2007**, *35*, W588–W594. [CrossRef]

70. Fukasawa, T.; Enomoto, A.; Miyagawa, K. Serine–Threonine Kinase 38 regulates CDC25A stability and the DNA damage-induced G2/M checkpoint. *Cell. Signal.* **2015**, *27*, 1569–1575. [CrossRef]

71. Lehtinen, M.K.; Yuan, Z.; Boag, P.R.; Yang, Y.; Villén, J.; Becker, E.B.; DiBacco, S.; de la Iglesia, N.; Gygi, S.; Blackwell, T.K.; et al. A conserved MST-FOXO signaling pathway mediates oxidative-stress responses and extends life span. *Cell* **2006**, *125*, 987–1001. [CrossRef]

72. Sanphui, P.; Biswas, S.C. FoxO3a is activated and executes neuron death via Bim in response to β-amyloid. *Cell Death Dis.* **2013**, *4*, e625. [CrossRef] [PubMed]

73. Valis, K.; Prochazka, L.; Boura, E.; Chladova, J.; Obsil, T.; Rohlena, J.; Truksa, J.; Dong, L.F.; Ralph, S.J.; Neuzil, J. Hippo/Mst stimulates transcription of the proapoptotic mediator *NOXA* in a FoxO1-dependent manner. *Cancer Res* **2011**, *71*, 946–954. [CrossRef] [PubMed]

74. Morimoto, N.; Nagai, M.; Miyazaki, K.; Kurata, T.; Takehisa, Y.; Ikeda, Y.; Kamiya, T.; Okazawa, H.; Abe, K. Progressive decrease in the level of YAPdeltaCs, prosurvival isoforms of YAP, in the spinal cord of transgenic mouse carrying a mutant *SOD1* gene. *J. Neurosci. Res.* **2009**, *87*, 928–936. [CrossRef]

75. He, Y.; Emoto, K.; Fang, X.; Ren, N.; Tian, X.; Jan, Y.N.; Adler, P.N.; Kurischko, C.; Kuravi, V.K.; Wannissorn, N.; et al. *Drosophila* Mob family proteins interact with the related tricornered (Trc) and warts (Wts) kinases. *Mol. Biol. Cell* **2005**, *16*, 4139–4152. [CrossRef] [PubMed]

76. Emoto, K.; He, Y.; Ye, B.; Grueber, W.B.; Adler, P.N.; Jan, L.Y.; Jan, Y.N. Control of dendritic branching and tiling by the Tricornered-kinase/Furry signaling pathway in Drosophila sensory neurons. *Cell* **2004**, *119*, 245–256. [CrossRef] [PubMed]

77. Norkett, R.; del Castillo, U.; Lu, W.; Gelfand, V.I. Ser/Thr kinase Trc controls neurite outgrowth in Drosophila by modulating microtubule-microtubule sliding. *eLife* **2020**, *9*, e52009. [CrossRef]

78. Wang, L.H.; Baker, N.E. Salvador-Warts-Hippo pathway regulates sensory organ development via caspase-dependent nonapoptotic signaling. *Cell Death Dis.* **2019**, *10*, 669. [CrossRef]

79. Lin, C.H.; Hsieh, M.; Fan, S.S. The promotion of neurite formation in Neuro2A cells by mouse Mob2 protein. *FEBS Lett.* **2011**, *585*, 523–530. [CrossRef]

80. Campbell, M.; Ganetzky, B. Identification of Mob2, a novel regulator of larval neuromuscular junction morphology, in natural populations of Drosophila melanogaster. *Genetics* **2013**, *195*, 915–926. [CrossRef]

81. O'neill, A.C.; Kyrousi, C.; Einsiedler, M.; Burtscher, I.; Drukker, M.; Markie, D.M.; Kirk, E.P.; Götz, M.; Robertson, S.P.; Cappello, S. Mob2 Insufficiency Disrupts Neuronal Migration in the Developing Cortex. *Front. Cell. Neurosci.* **2018**, *12*, 57. [CrossRef] [PubMed]

82. Hondius, D.C.; Eigenhuis, K.N.; Morrema, T.H.J.; van der Schors, R.C.; van Nierop, P.; Bugiani, M.; Li, K.W.; Hoozemans, J.J.M.; Smit, A.B.; Rozemuller, A.J.M. Proteomics analysis identifies new markers associated with capillary cerebral amyloid angiopathy in Alzheimer's disease. *Acta Neuropathol. Commun.* **2018**, *6*, 46. [CrossRef] [PubMed]

83. Satoh, J.I.; Kawana, N.; Yamamoto, Y. Pathway Analysis of ChIP-Seq-Based NRF1 Target Genes Suggests a Logical Hypothesis of their Involvement in the Pathogenesis of Neurodegenerative Diseases. *Gene Regul. Syst. Biol.* **2013**, *7*, 139–152. [CrossRef] [PubMed]

84. McInnes, J. Insights on altered mitochondrial function and dynamics in the pathogenesis of neurodegeneration. *Transl. Neurodegener.* **2013**, *2*, 12. [CrossRef]

85. Oláh, J.; Vincze, O.; Virók, D.; Simon, D.; Bozsó, Z.; Tőkési, N.; Horváth, I.; Hlavanda, E.; Kovács, J.; Magyar, A.; et al. Interactions of pathological hallmark proteins: Tubulin polymerization promoting protein/p25, beta-amyloid, and alpha-synuclein. *J. Biol. Chem.* **2011**, *286*, 34088–34100. [CrossRef]

86. Hidalgo, A.; Ffrench-Constant, C. The control of cell number during central nervous system development in flies and mice. *Mech. Dev.* **2003**, *120*, 1311–1325. [CrossRef]

87. Guo, Y.; Zeng, Q.; Brooks, D.; Geisbrecht, E.R. A conserved STRIPAK complex is required for autophagy in muscle tissue. *Mol. Biol. Cell* **2023**, *34*, ar91. [CrossRef]

Article

Lab-It Is Taking Molecular Genetics to School

Márcio Simão [1,2,*], Natércia Conceição [1,2,3], Susana Imaginário [4], João Amaro [5] and Maria Leonor Cancela [1,2,3]

[1] Comparative, Adaptive and Functional Skeletal Biology (BIOSKEL) Lab, Centre of Marine Sciences (CCMAR), Universidade do Algarve, 8005-135 Faro, Portugal; nconcei@ualg.pt (N.C.); lcancela@ualg.pt (M.L.C.)
[2] Faculty of Medicine and Biomedical Sciences, Universidade do Algarve, 8005-135 Faro, Portugal
[3] Algarve Biomedical Center (ABC), Universidade do Algarve, 8005-135 Faro, Portugal
[4] Divisão de Empreendedorismo e Transferência de Tecnologia (CRIA), Universidade do Algarve, 8005-135 Faro, Portugal; ssimaginario@ualg.pt
[5] Innovatio Sensum Consultoria Sociedade Unipessoal LDA, Rua Padre António Macedo, n° 16, 7540-200 Santiago do Cacém, Portugal; riosecoamaro@hotmail.com
* Correspondence: masimao@ualg.pt

Abstract: The Molecular Genetics Mobile Lab or "Laboratório itinerante de Genética Molecular" (Lab-it) was funded in 2008 by Leonor Cancela to promote the learning of molecular genetics which had been introduced at that time into high school biology programms. The project aimed to introduce hands-on laboratory activities in molecular genetics to complement the theoretical concepts taught in school. These included the development of experimental protocols based on theoretical scenarios focusing on themes of forensics sciences, biomedical applications, diagnostic methods, and ecological research using basic molecular biology techniques, such as DNA extraction, polymerase chain reaction (PCR), electrophoresis, and restriction enzyme application. In these scenarios, the students execute all the procedures with the help of the Lab-it instructor and using the Lab-it equipment, followed by a discussion of the results with all the participants and the school teacher. These approaches help the students to consolidate the concepts of molecular biology and simultaneously promote discussions on new advances in the area and choices for university careers. In addition to practical sessions, Lab-it also promotes seminars on topics of interest to the students and teachers. Since 2008, 18 high schools have participated in the region of Algarve, averaging each year about 400 students participating in practical activities. In 2021, despite the COVID pandemic, 9 schools and 379 students were involved in Lab-it practical sessions and 99% of them considered the activity to contribute to better understanding the molecular biology methods approached in theoretical classes and expressed high interest in those sessions.

Keywords: education; molecular genetics; lab-it; laboratory hands-on sessions

Citation: Simão, M.; Conceição, N.; Imaginário, S.; Amaro, J.; Cancela, M.L. Lab-It Is Taking Molecular Genetics to School. *BioChem* **2022**, 2, 160–170. https://doi.org/10.3390/biochem2020011

Academic Editor: Buyong Ma

Received: 15 January 2022
Accepted: 2 June 2022
Published: 9 June 2022

Publisher's Note: MDPI stays neutral with regard to jurisdictional claims in published maps and institutional affiliations.

1. Introduction

With the introduction of biochemistry and molecular biology concepts in biology high school teaching programs all over the world [1–4] and particularly in Portugal, teachers have been reporting difficulties in teaching students the concepts associated with the flow of genetic information in cells and the molecular biology methods currently used for biotechnology applications [5–8]. Students have reported difficulties in understanding the correlation of biomolecules like deoxyribonucleic acid (DNA), messenger ribonucleic acid (mRNA), and proteins with cell functions, and basic concepts like genes, proteins, and genetic phenotypes are often not fully understood [6,8]. Biochemistry and molecular biology concepts are somehow abstract because they occur at the molecular level of biological systems. Available information indicates that students find problems to understand those concepts due to the degree of complexity and poor contextualization related to the flow of genetic information due to the increased number of concepts approached by the scientific program of biology classes and the time available to students to learn it [4,6,7]. In addition, misconceptions related to the teacher's preparation and lower degree of training updates were identified as contributing factors [1,7,8]. To complement formal learning taking place within the framework of classes and based on the description of theoretical concepts by the teacher, the introduction of practical projects associated with these subjects has proven to help students understand their complexity [8–10].

To optimize the teaching of molecular biology concepts included in the high school biology program in Portugal, several teachers started to contact the universities to help them refresh their know-how in those topics. Attending to this, in 2008, Leonor Cancela, professor of molecular genetics at the University of Algarve, started the project called The Molecular Genetics Mobile Lab or "Laboratório itinerante de Genética Molecular" (Lab-it) to help schools and biology teachers in the Algarve region to explain molecular biology concepts to their students. Attending that the equipment necessary to execute molecular biology experiments was not accessible to schools, the idea was to equip an itinerary laboratory, the Lab-it, with the equipment and reagents necessary to execute molecular genetics experiments in loco, in each participating school. The Lab-it could then visit every high school in Algarve to allow biology students to execute the experiments based on hands-on molecular biology techniques like DNA extraction, PCR, electrophoresis, and enzymatic restriction analysis of DNA fragments. The experimental procedures were contextualized in four theoretical scenarios, namely (i) criminal forensic investigation, (ii) prenatal genetic screenings, (iii) molecular detection and diagnostic of patients infected with bacteria carrying multi-resistance to antibiotics, and (iv) molecular identification of bacteria and fungi species from environmental samples. This study presents the results obtained from high school students that participated in Lab-it sessions in the year 2021. The aim was to understand in which measure the Lab-it activities developed by the students contributed to improve (i) their comprehension of molecular genetics principles, (ii) their level of success in the implementation of hands-on protocols by the students, and (iii) in which measure the Lab-it activities contributed to motivating students to continue their academic training and apply to university, in particular, to follow molecular genetics and biochemistry areas.

2. Materials and Methods

2.1. Educational Contextualization of Students

The Lab-it project develops its activities in the Algarve region, in the south of Portugal. In 2021, the project was implemented in 9 of the 18 high schools contacted. Although more schools were interested in collaborating, the COVID-19 pandemic imposed restrictions on some schools, which prevented them from joining the project. A total of 379 students participated in 33 Lab-it sessions. Portugal's education system is organized into three stages, basic, primary, and secondary education and the students that took part in the practical sessions belonged to the secondary education stage, namely the last two years of high school, 11th and 12th grades. All the students had chosen to study biology when entering the secondary education level.

2.2. Methodology of Practical Session

The Lab-it practical sessions were organized with a maximum of 16 students each, with a duration of 3 h. Each activity starts with a presentation of molecular genetics concepts associated with the flow of genetic information within the cells, discussing DNA hereditary properties, three-dimensional structure, composition, and information on genetic hereditary transmission mechanisms. Additionally, students were also introduced to basic concepts associated with methods like DNA extraction, polymerase chain reaction (PCR), electrophoresis, restriction enzyme analysis, and interpretation of results. Students were provided with examples of how to interpret results from a PCR after electrophoresis, contextualizing the polymorphisms evaluated (homozygotes/heterozygotes or presence-absence of bacteria and fungi by ribosomal gene detection) and molecular context of the target gene amplified for each scenario. After the presentation, students learn the composition of a PCR reaction and correlate it with DNA replication mechanisms. This is followed by learning how to use micropipettes. After that, each group of students assembled the PCR reaction in a tube (a maximum of four groups with four elements each) and applied it to the thermocycler for amplification. All the practical steps were executed by the students after the initial explanation by the Lab-it organizer. After a small break, the principles of electrophoresis were discussed with the students as well as the equipment to be used. Correlation between the interpretation of the results discussed at the beginning and the fundaments of the technique were approached at this point. The students prepared the samples and applied them to an agarose gel for electrophoresis. During the time that electrophoresis was running, we discussed with the students any doubts regarding the practical sessions, new technical approaches used in research, and options for university courses. After electrophoresis, the agarose gel was observed in an Ultraviolet transilluminator, and the results were discussed. Altogether, this process allows students to contextualize experimental approaches and observe the application of these methods to answer the questions in the scenarios chosen (Table 1).

Table 1. Description of Lab-it scenarios applied during practical sessions.

Scenario	Title	Techniques Used	Objectives
1	Who committed the crime?	Pipetting with micropipettes, Centrifugation, PCR, electrophoresis	Amplification of polymorphisms in two genes (*PLAT* and *AMELX)* to contribute to genetic profile of a crime suspect
2	Where are the *Pseudomonas*? A case of Multi-resistance infection in Algarve	Pipetting with micropipettes, Centrifugation, PCR, electrophoresis, restriction enzyme cut selection	Amplification of *16S* ribosomal subunit fragment by PCR from samples of patients to identify *Pseudomonas* sp. polymorphism with *EcoRI* restriction enzyme analysis
3	Prenatal screening analysis for hereditary disease and gender determination	Pipetting with micropipettes, Centrifugation, PCR, electrophoresis	Screening for the presence of *Alu* fragment insertion in *CFTR* gene and gender determination through amplification of *AMELX* fragment
4	Microbial ecology: Looking for bacteria and fungi	Pipetting with micropipettes, Centrifugation, PCR, electrophoresis, restriction enzyme cut selection	Amplification of DNA fragments of *16S* and *18S* (specific for fungal species) subunits to confirm presence and absence within the ecological niche and *EcoRI* restriction analysis for fragments amplified

The implementation of molecular biology techniques is included as part of four distinct scenarios (Table 1). These scenarios allow students to assimilate the molecular biology principles within a practical application example.

For human biology scenarios, the students amplify two polymorphisms common in the human genome that should favor the understanding of concepts like DNA heredity of polymorphisms and consequences to the homozygotic and heterozygotic state of the human genome analyzed. The first polymorphism that we have used was a common Alu insertion in an intron of the gene that encodes for plasminogen activator, tissue type (*PLAT*) [11]. This was used in scenarios 1 and 3 (Table 1). In the first scenario, it was used in the context of a polymorphism specific to an individual, which could be involved in the identification of suspects at a crime scene [11]. For the third scenario, this polymorphism was used as a theoretical example for a possible *Alu* insertion into an exon of the gene cystic fibrosis transmembrane conductance regulator (*CFTR*), as described previously [12], but based on the amplification of *Alu* insertion in gene *PLAT*. Mutations in this gene are responsible for the majority of cystic fibrosis cases in the world [13], and this context allowed students to think about the hereditary recessive polymorphisms that can lead to diseases. Additionally, for scenarios 1 and 3 (Table 1), we have used primers specific to the Amelogenin X-linked gene (*AMELX*), which allowed the identification of chromosomes X and Y [13] and could be used in scenario 1 for gender identification of the alleged suspect and scenario 3 for prenatal gender determination.

For scenarios 2 and 4 (Table 1), the students amplified fragments of 16S ribosomal RNA (*16S*) using primers (8F and 1542R) for molecular detection of bacteria as described by Galkiewicz and Kellogg (2008) [14]. The context of scenario 2 theoretical objective was to detect bacteria in patients suspected of being infected by a multi-resistance strain of *Pseudomonas* sp. The students used samples of DNA isolated from oral epithelium cells, which allowed the detection of bacteria. Additionally, we have analyzed several *16S* sequences of several bacteria strains within the fragment amplified by primers (8F and 1542R) [14] and identified an *EcoRI* restriction enzyme recognition site (5′-GAATTC-3′) in most oral epithelium common bacteria strains, producing two fragments (840bp and 660bp). It was also possible to observe that *Lactobacillus* sp. and *Pseudomonas* sp. do not have this *EcoRI* recognition site. This fact allowed the students to do a restriction analysis for possible detection of *Pseudomonas* sp. as a theoretical multi-resistance strain and because *Lactobacillus* sp. are easy to detect in the oral epithelium [12], this assay could be easily used to simulate scenario 2 application (Table 1). This allowed students to understand how to differentiate bacterial strains using a restriction enzyme digestion.

The same principle was used for bacterial detection in scenario 4, but in this context, the main objective was to detect bacteria strains from different ecological niches. Because the *EcoRI* restriction enzyme recognition site is absent in *Lactobacillus* sp. and *Pseudomonas* sp. we can promote the discussion with the students for differentiation of these strains in theoretical niches of scenario 4. Additionally, we used the primers nu-SSU-0817-5′ and nu-SSU-1536-3′ for 18S ribosomal RNA (*18S*) genes fragment amplification as described by Borneman and Hartin (2000) [15]. The amplification of fungal microbial species within the context of scenario 4 allowed students to think about microbial diversity, prokaryote, and eukaryote, in several niches and the potential of molecular biology techniques for detection and characterization of these ecological niches.

2.3. Data Collection and Statistical Analysis

The inquiry that was prepared for the students was composed of 19 questions. The questions were subdivided into four sections, the first one (two questions) asking about age and high school identification, the second (nine questions) related to the execution and opinion about the activities and methods developed by the students, the third (five questions), related with options of the students concerning possible choices of university courses, and the fourth section (two questions) was about the impact of COVID-19 disease on their motivation to understand nucleic acid detection methods and interest in biological sciences. The inquiry was available on an online platform (https://www.survio.com/pt/, accessed on 31 December 2021). The access to the inquiry was done through a site address or quick response code (QR code) and was answered by the students at the end of each practical session. The inquiry was anonymous. From the 19 questions asked, in this paper, we present the results obtained for 15 questions. The answers presented are essentially related to the practical sessions attended by the students, options for university courses, and the impact of the COVID-19 disease.

For this study, we applied a univariant analysis for which we have divided the answers into two types, classification from 1–10 and parts of a whole. For the answers requiring a classification ranging from 1 to10, data was presented through a box and violin graphics, allowing distribution of answers analysis, and reported along with the average classification, coefficients of variation (determined by the relative standard deviation) [16], and respective confidence intervals (CI) at 95%. The parts of a whole analysis were represented by pie charts or bar graphics and results were reported as percentage or number of answers to each category. The confidence interval at 95% was presented along with the graphics presented. The CI at 95% was calculated based on the hybrid Wilson/Brown method [17].

3. Results

In 2021, Lab-it visited a total of nine high schools in the Algarve region, south of Portugal, and a total of 379 students participated in 33 practical sessions. Of the 379 students that took part in those sessions, 367 answered the inquiry. The student's ages ranged from 16 to 20 years, with 42.0% $\geq$ 18, 54.8% = 17, and 3.3% = 16 years old, respectively.

The students were asked to classify from 1–10 how (i) interesting and (ii) motivating were the practical sessions. The students considered the session very interesting with a mean score of X = 9.10 (CV = 12.8%) and motivating X = 8.8 (CV = 16.0%), (Figure 1). When asked about the duration of the practical session (3 h) 76.6% considered that was adequate, 11.4% would like more time and 12.0% preferred less time (Figure 1). Regarding Lab-it practical implementation 69.6% and 29.8% considered "very good" and "good", respectively (Figure 1).

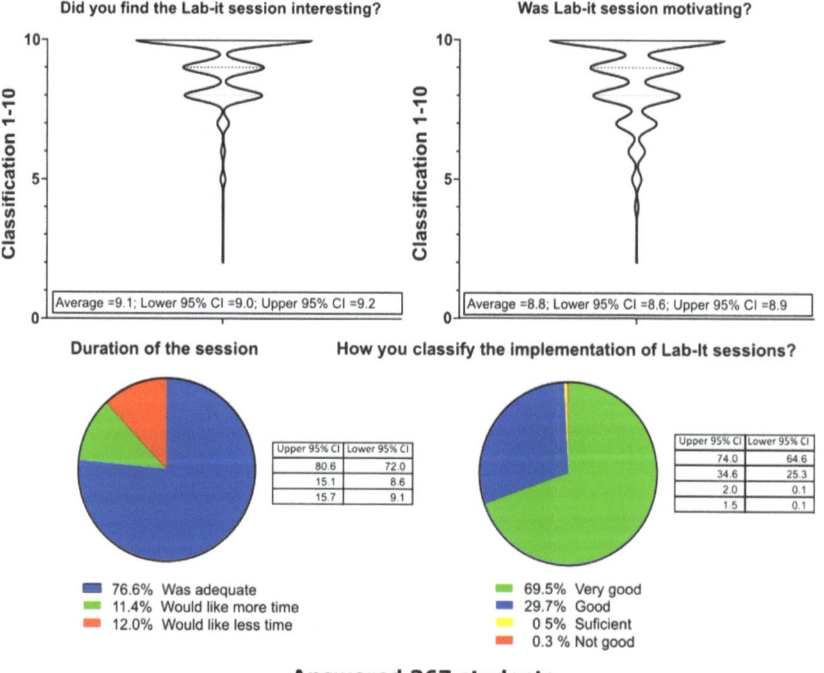

Figure 1. Opinion of students regarding the application of practical activities of Lab-it.

Regarding the most positive aspects of the practical sessions, the preferred option (292 answers) was the opportunity to have a hands-on experience with Lab-it equipment, followed by the application in experiments that related to themes studied in biology classes (200 answers) (Figure 2). Students could answer more than one option or write their option if they wanted to add any comments. In addition, it was asked if students would change anything about the practical session. Results showed that 218 students did not change anything and 133 would like to apply more techniques (Figure 2).

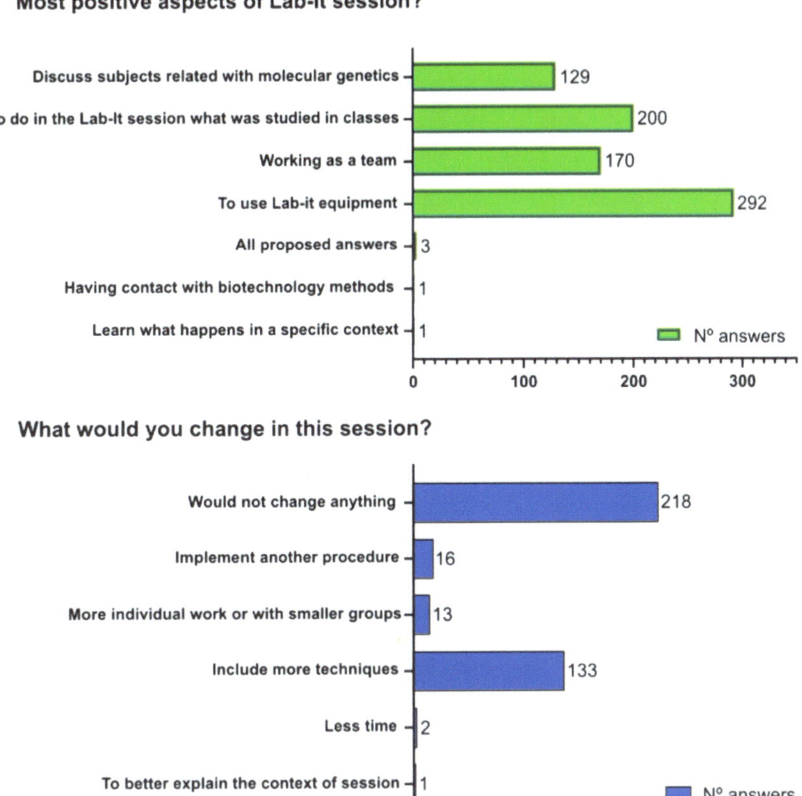

Figure 2. Compilation of opinions about what were the most positive aspects of Lab-it activities and what would students change in the sessions.

When inquiring the students if what was applied in the scenarios and techniques used in the practical session were in line with the biology discipline program, 73.3% considered "very good" and 24,0% "good" (Figure 3). Only 2.7% answered that it was "sufficient ", and none of them choose "not good (Figure 3). In addition, 99.2% considered that the practical sessions contributed to a better understanding of the techniques and methods of molecular biology applied (Figure 3), and 98.9% of the students believed it would be important to do more practical activities like Lab-it sessions to better understand class subjects given in high schools (Figure 3).

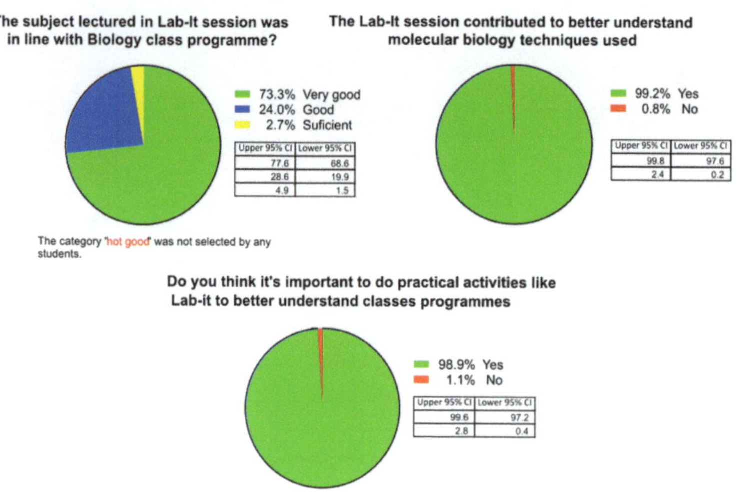

Figure 3. Students' opinions about the correlation between the practical application of molecular biology methods and theoretical subject approaches in biology classes.

Another question concerned the student's options of continuing to university and choice of courses. 89.9% of the students indicated they were planning to apply to the university (Figure 4). Of those, 40.6% would apply to biological sciences courses and 31.3% were considering going to a course in this area (Figure 4). About 28.1% did not plan to apply to a course in this area (Figure 4). Regarding the students that answered yes or maybe to the question of continuing into biological sciences, it was asked to classify from 1-to 10 how the Lab-it session contributed to their choice and on average answered X = 7.8 (CV = 24.8%), (Figure 4).

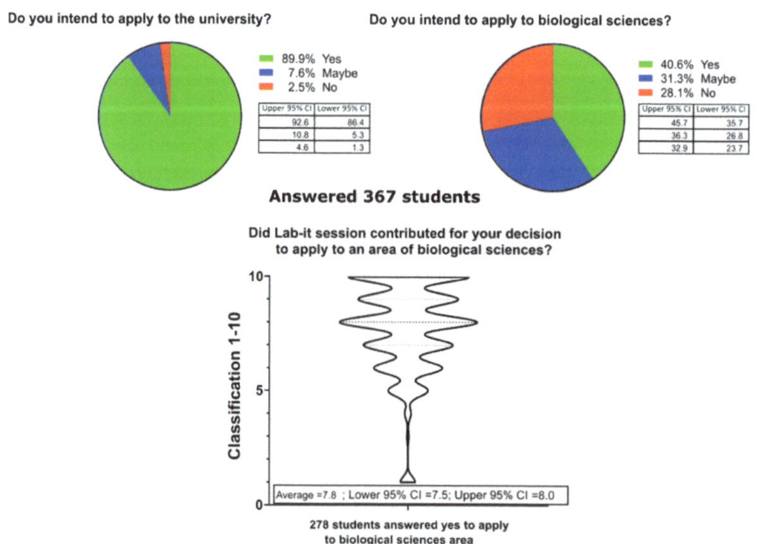

Figure 4. Understanding if students want to proceed with studies in the university and if the Lab-it session could have contributed to students' choice towards going into a biological sciences area.

Attending that the Lab-it practical sessions were done following severe COVID-19 restrictions, it was asked to students to classify from 1–10 if the SARS-CoV-2 pandemic changed the way that they studied, on average answered X = 7.7 (CV = 29.8%), (Figure 5). It was also asked if the COVID-19 pandemic increased their interest in the biomedical sciences area and students classified as X = 6.0 (CV = 47.7%), (Figure 5).

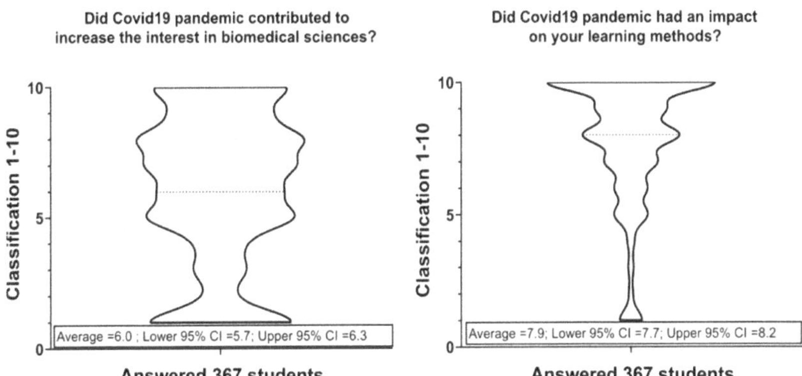

Figure 5. Impact of COVID-19 pandemic on students' interest in biomedical sciences and habits of study.

4. Discussion

The Lab-it project aimed to contribute to improving the teaching of molecular biology and biochemistry concepts to high school students in the region of Algarve, south of Portugal. The objective of the present study was to understand if the technical approaches done in Lab-it sessions within specific scenarios contributed in a significant way to increase the understanding of the molecular concepts discussed in the high school biology program, from the point of view of the students. For that, an inquiry was prepared with questions focused on specific items that could allow us to understand if the Lab-it sessions had contributed to fulfilling the proposed objectives.

The first two questions were done with the objective of understanding if the activity was appealing and could bring attention to molecular genetics concepts. Results suggested an increased interest and motivation to learn concepts developed in the practical activity (Figure 1), contributing to improve motivation and interest, as described before when students attended practical classes [18–20]. In addition, it was asked to students about the application and duration of Lab-it sessions and student opinions were highly positive as represented in Figure 1. The most significant results of this inquiry are related to the fact that students believe that the Lab-it practical session to be an extremely positive contribution to understanding the molecular biology and would like to have similar activities for other subjects (Figure 3), in agreement with what was previously believed about the importance of practical work in molecular genetics and sciences in general [9,10,21]. These results are in agreement with previously published studies which showed that complementing the teaching of theoretical concepts with practical methods can be a significant contribution to students' comprehension of molecular genetics and biochemistry concepts [18,19,22,23]. In line with this, the opinion of students confirmed that practical activities were a positive aspect to understand molecular genetics. When asked about the most positive aspects of the practical session, students indicated to be the possibility of hands-on use of Lab-it equipment and be able to apply concepts thought in biological theoretical classes to answer specific questions (Figure 2). In addition, when asked which aspects they would change in the Lab-it activity, the majority answered that they would like to include the use of more techniques (Figure 2) [18,19] thus confirming their interest in hands-on experiments.

The analysis of the possibility to apply to university courses showed us that the vast majority of students were planning to continue their studies into the university (89.9%) and of this, 40.6% were planning to apply to a course in biological sciences (Figure 4). These results showed that the student population which attends science and technology courses at the upper secondary education level is highly motivated to proceed to study at the university. In addition, students that would like to go to biological sciences course or were thinking about it were asked to classify from 1–10 what was the impact of Lab-it activity in their decision. The students on average classified as 7.8 with a CV of 24.8%, suggesting that most students classified >5 (Figure 4), indicating that Lab-it activities had a positive impact in motivating them to continue their studies in this area.

The Lab-it practical sessions were done during the COVID-19 pandemic after restrictions to presential classes were lifted. Attending to that, we asked students to classify from 1–10 how this context affected studying habits, and on average students classified as $X = 7.9$ (Figure 5), suggesting a significant impact of the SARS-CoV-2 pandemic in schools. These results are in agreement with what was published before [24,25]. Students were also asked to classify if this pandemic context led to an increase in their interest in biomedical sciences and students on average classified as 6.0 with a CV of 47.7%, suggesting that their opinion regarding this question was dispersed and not consensual (Figure 5).

This study describes, from the students' point of view, the impact of Lab-it sessions on comprehension of molecular genetics concepts and clearly shows how these activities can motivate students and contribute to a better understanding of molecular genetics and biochemistry concepts. However, this study had some limitations, in particular by lacking the possibility of a follow-up of the students after participation in Lab-it to evaluate its impact on the continuation of their training. We propose in the near future to follow some of the students that participated in Lab-it sessions and evaluate how their comprehension of basic molecular genetics concepts contributes to their success in the final exams and facilitates their integration into the university courses in this area and compare it with students that did not benefit from similar activities. In addition, it would be important to inquire from the high school teachers that collaborated with Lab-it about the differences found in student's progress and motivation before and after the participation in the Lab-it project. It will be necessary to develop a practical method to quantify this impact. This study would also benefit from a compilation of results of inquiries to the teachers that collaborated with Lab-it. This is planned for future studies.

5. Conclusions

The main objective of this study was to understand, from the student's point of view, the importance of Lab-it practical sessions implemented for their scientific training, and to obtain evidence of the positive points and changes that could be done to improve it. We can conclude with these results that students have a highly positive impression of the execution of practical activities, with a focus on learning to use Lab-it equipment to better understand the technology involved and apply this know-how to solve specific questions through hands-on experiences. They even suggested the inclusion of additional techniques. It was important to understand also if the model of those sessions was contributing to increasing the understanding of molecular genetics and biochemistry concepts approached in biology classes and we can conclude that students considered it highly relevant and important to better understand the subjects taught in their biology program. Results also showed that the students were highly motivated to apply to a university and a considerable percentage found that participating in Lab-it practical sessions help them to decide to choose an area of biological sciences. This study represents an initial step to a follow-up study about the impact of Lab-it activities in high schools and how they could influence students' learning skills and choices for progressing into university studies.

Author Contributions: Conceptualization: M.S. and M.L.C. Experimental design: M.S. Application of Lab-it: M.S., N.C., S.I., J.A. and M.L.C. Writing: M.S. and M.L.C. Review & editing: M.S., N.C., S.I., J.A. and M.L.C. All authors have read and agreed to the published version of the manuscript.

Funding: The Lab-it is part of KCITAR, a project funded by a consortium of entities: Algarve STP-Algarve Systems and Technology Partnership, Portugal 2020 program, ESF-European Social Fund and University of Algarve (GRANT_NUMBER: ALG-08-5864-FSE-000004 K.CITAR/10IS00018). Lab-It collaborated with the BIOSKEL laboratory and Centre of Marine Sciences (CCMAR), University of Algarve, Faro, Portugal, funded by FCT—Foundation for Science and Technology-through project UIDB/04326/2020.

Institutional Review Board Statement: Not applicable.

Informed Consent Statement: Informed consent was obtained from all subjects involved in the study.

Data Availability Statement: Full data regarding inquiries answers are available upon request to authors.

Acknowledgments: We want to acknowledge all the students, teachers and high schools ("Escola Secundária Vila Real de Santo António", "Escola Secundária Tomás Cabreira (Faro)", "Escola Secundária Pinheiro e Rosa (Faro)", "Escola Secundária de Loulé", "Escola Secundária Júlio Dantas (Lagos)", "Escola Secundária José Belchior Viegas (São Brás de Alportel)", "Escola Secundária João de Deus (Faro)", "Escola Secundária Francisco Fernandes Lopes (Olhão)" and "Colégio Nobel (Lagoa))", that have participated in Lab-it sessions in the year 2021.

Conflicts of Interest: The authors have no conflict of interest do disclose.

References

1. Ozcan, T.; Ozgur, S.; Kat, A.; Elgun, S. Identifiying and Comparing the Degree of Difficulties Biology Subjects by Adjusting it is Reasons in Elemantary and Secondary Education. *Procedia-Soc. Behav. Sci.* **2014**, *116*, 113–122. [CrossRef]
2. Wood, E.J. Biochemistry and molecular biology teaching over the past 50 years. *Nat. Rev. Mol. Cell Biol.* **2001**, *2*, 217–221. [CrossRef] [PubMed]
3. Rotbain, Y.; Marbach-Ad, G.; Stavy, R. The effect of different molecular models on high school students' conceptions of molecular genetics. *Sci. Educ. Rev.* **2008**, *7*, 59–64. Available online: https://files.eric.ed.gov/fulltext/EJ1050891.pdf (accessed on 31 December 2021).
4. Machová, M.; Ehler, E. Secondary school students' misconceptions in genetics: Origins and solutions. *J. Biol. Educ.* **2021**, 1–14. [CrossRef]
5. Saka, A.; Cerrah, L.; Akdeniz, A.R.; Ayas, A. A cross-age study of the understanding of three genetic concepts: How do they image the gene, DNA and chromosome? *J. Sci. Educ. Technol.* **2006**, *15*, 192–202. [CrossRef]
6. Eklund, J.; Rogat, A.; Alozie, N.; Krajcik, J. Promoting student scientific literacy of molecular genetics and genomics. In Proceedings of the Annual Meeting of the National Association for Research in Science Teaching, New Orleans, LA, USA, April 2007; pp. 1–20. Available online: http://websites.umich.edu/~{}hiceweb/PDFs/2007/Genetics_NARST_07.pdf (accessed on 30 December 2021).
7. Marbach-Ad, G. Attempting to break the code in student comprehension of genetic concepts. *J. Biol. Educ.* **2001**, *35*, 183–189. [CrossRef]
8. Orhan, T.Y.; Sahin, N. The impact of innovative teaching approaches on biotechnology knowledge and laboratory experiences of science teachers. *Educ. Sci.* **2018**, *8*, 213. [CrossRef]
9. Katajavuori, N.; Lindblom-Ylänne, S.; Hirvonen, J. The significance of practical training in linking theoretical studies with practice. *High. Educ.* **2006**, *51*, 439–464. [CrossRef]
10. Zimmerman Nilsson, M.H. Practical and theoretical knowledge in contrast: Teacher educators discursive positions. *Aust. J. Teach. Educ.* **2017**, *42*, 29–42. [CrossRef]
11. Perna, N.T.; Batzer, M.A.; Deininger, P.L.; Stoneking, M. Alu insertion polymorphism: A new type of marker for human population studies. *Hum. Biol.* **1992**, *64*, 641–648. Available online: https://www.jstor.org/stable/41465107 (accessed on 31 December 2021).
12. Caufield, P.W.; Schön, C.N.; Saraithong, P.; Li, Y.; Argimón, S. Oral Lactobacilli and Dental Caries: A Model for Niche Adaptation in Humans. *J. Dent. Res.* **2015**, *94*, 110S–118S. [CrossRef] [PubMed]
13. Mall, M.A.; Hartl, D. CFTR: Cystic fibrosis and beyond. *Eur. Respir. J.* **2014**, *44*, 1042–1054. [CrossRef] [PubMed]
14. Galkiewicz, J.P.; Kellogg, C.A. Cross-kingdom amplification using Bacteria-specific primers: Complications for studies of coral microbial ecology. *Appl. Environ. Microbiol.* **2008**, *74*, 7828–7831. [CrossRef]
15. Borneman, J.; Hartin, R.J. PCR primers that amplify fungal rRNA genes from environmental samples. *Appl. Environ. Microbiol.* **2000**, *66*, 4356–4360. [CrossRef] [PubMed]
16. Gao, Y.; Ierapetritou, M.G.; Muzzio, F.J. Determination of the confidence interval of the relative standard deviation using convolution. *J. Pharm. Innov.* **2013**, *8*, 72–82. [CrossRef]
17. Tsai, W.Y.; Chi, Y.; Chen, C.M. Interval estimation of binomial proportion in clinical trials with a two-stage design. *Stat. Med.* **2008**, *27*, 15–35. [CrossRef]

18. Hofstein, A.; Mamlok-naaman, R. The laboratory in science education: The state of the art. *Chem. Educ. Res. Pract.* **2007**, *8*, 105–107. [CrossRef]
19. Holfstein, A.; Lunetta, V.N. The Role of the Laboratory in Science Teaching: Neglected Aspects of Research. *Am. Educ. Res. Assoc.* **1982**, *52*, 201–217. [CrossRef]
20. Novelli, E.L.B.; Fernandes, A.A.H. Students' preferred teaching techniques for biochemistry in biomedicine and medicine courses. *Biochem. Mol. Biol. Educ.* **2007**, *35*, 263–266. [CrossRef]
21. Millar, R. *The Role of Practical Work in the Teaching and Learning of Science*; Paper prepared for the Committee; High School Science Laboratories: Role and Vision, National Academy of Sciences: Washington, DC, USA, 2004; pp. 1–24. Available online: https://sites.nationalacademies.org/cs/groups/dbassesite/documents/webpage/dbasse_073330.pdf (accessed on 31 December 2021).
22. Scott, P.H.; Veitch, N.J.; Gadegaard, H.; Mughal, M.; Norman, G.; Welsh, M. Enhancing theoretical understanding of a practical biology course using active and self-directed learning strategies. *J. Biol. Educ.* **2017**, *52*, 184–195. [CrossRef]
23. Leask, R.; Cronje, T.; Holm, D.E.; van Ryneveld, L. The impact of practical experience on theoretical knowledge at different cognitive levels. *J. S. Afr. Vet. Assoc.* **2020**, *91*, e1–e7. [CrossRef] [PubMed]
24. Pokhrel, S.; Chhetri, R. A Literature Review on Impact of COVID-19 Pandemic on Teaching and Learning. *High. Educ. Futur.* **2021**, *8*, 133–141. [CrossRef]
25. Basilaia, G.; Kvavadze, D. Transition to Online Education in Schools during a SARS-CoV-2 Coronavirus (COVID-19) Pandemic in Georgia. *Pedagog. Res.* **2020**, *5*, 1–9. [CrossRef]

Article

cEpiderm, a Canine Skin Analog Suitable for In Vivo Testing Replacement

Mariana Marques [1], João Nunes [2], Bárbara Ustymenko [1], Luísa Fialho [3], Luís Martins [3,4], Anthony J. Burke [5,6], Cesar Filho [7], Alexandre C. Craveiro [7], Ana R. Costa [1,8,9,*], Sandra Branco [3,4] and Célia M. Antunes [1,8,9,*]

[1] Institute of Earth Sciences—ICT, University of Évora, 7000-671 Évora, Portugal
[2] Department of Chemistry, School of Sciences and Technology, University of Évora, 7000-671 Évora, Portugal
[3] Department of Veterinary Medicine, School of Sciences and Technology, University of Évora, 7002-554 Évora, Portugal
[4] Mediterranean Institute for Agriculture, Environment and Development—MED, University of Évora, 7002-554 Évora, Portugal
[5] Pharmacuetical Chemistry Section, Faculty of Pharmacy, Polo das Ciências da Saúde, University of Coimbra, 3000-548 Coimbra, Portugal
[6] Associated Laboratory for Green Chemistry of the Network of Chemistry and Technology—LAVQ-REQUIMTE, University of Évora, 7000-671 Évora, Portugal
[7] BrInova—Bioquímica Lda, 7005-485 Évora, Portugal
[8] Department of Medical and Health Sciences, School of Health and Human Development, University of Évora, 7000-671 Évora, Portugal
[9] Centro Académico Clínico do Alentejo (C-TRAIL), 7000-671 Évora, Portugal
[*] Correspondence: acrc@uevora.pt (A.R.C.); cmma@uevora.pt (C.M.A.)

Citation: Marques, M.; Nunes, J.; Ustymenko, B.; Fialho, L.; Martins, L.; Burke, A.J.; Filho, C.; Craveiro, A.C.; Costa, A.R.; Branco, S.; et al. cEpiderm, a Canine Skin Analog Suitable for In Vivo Testing Replacement. *BioChem* **2022**, 2, 215–220. https://doi.org/10.3390/biochem2040015

Academic Editor: Buyong Ma

Received: 2 August 2022
Accepted: 12 October 2022
Published: 20 October 2022

Publisher's Note: MDPI stays neutral with regard to jurisdictional claims in published maps and institutional affiliations.

Abstract: Skin is one of the organs most tested for toxicity and safety evaluation during the process of drug research and development and in the past has usually been performed in vivo using animals. Over the last few years, non-animal alternatives have been developed and validated epidermis models for human and rat skin are already available. Our goal was to develop a histotypical canine skin analog, suitable for non-animal biocompatibility and biosafety assessment. Canine keratinocytes were seeded in an air-lift culture using an adapted version of the CELLnTEC protocol. Corrosion and irritation protocols were adapted from human EpiSkin™. For histological analysis, sample biopsies were fixed in neutral-buffered formalin, and paraffin slices were routinely processed and stained with hematoxylin and eosin. A canine multilayer and stratified epidermal-like tissue (cEpiderm), confirmed by histological analysis, was obtained. The cEpiderm tissue exhibited normal morphological and functional characteristics of epidermis, namely impermeability and an adequate response to stressors. The cEpiderm is a promising canine skin model for non-animal safety testing of veterinary pharmaceuticals and/or cosmetics, significantly contributing to reducing undesirable in vivo approaches. cEpiderm is therefore a valid canine skin model and may be made commercially available either as a service or as a product.

Keywords: histotypical skin culture; skin analog; skin; epidermis; non-animal testing

1. Introduction

Skin analogs or skin equivalents are extensively used in research and the testing of various products in different industries such as cosmetics, pharmaceuticals and skin care companies [1]. These analogs can also have applications in the treatment of acute and chronic wounds such as burns or diabetic wounds, additionally they aid in the research of new ways of treatments for various diseases such as melanoma and psoriasis [2,3].

Skin analogs have become quite popular because they are a sustainable and practical alternative compared to the use of live animals or mammalian skin explants, and they can also reduce significant errors and inaccuracies principally with regard to differences between species since the biochemistry and physiology of the skin varies from species to species [4].

Initially in vitro skin analogs consisted of a two-dimensional layer of cultured cells, mainly keratinocytes to represent the epidermal and outer layer of the skin, or fibroblasts that represented the dermal layer and were used mainly for toxicity assessment. However, these skin models lack the ability to mimic the penetration and absorption of different chemicals and materials of the live skin [5]. The development of new technologies and research has increased, especially in the case of human epidermal models capable of mimicking the in vivo skin, its morphology, its biomolecular and metabolic properties [6]. Skin analogs based on air-lift organotypic cultures have been developed, creating a three-dimensional skin model capable of simulating the permeability and absorption capacity of the in vivo skin. These 3D models may also incorporate components of a dermal layer, such as fibroblasts, or other skin cells, introducing complexity and versatility to these biosystems.

The skin barrier function is very complex and difficult to be artificially replicated because of the lack of skin equivalents including appendages, vasculature and lipids, thus the gold standard is still the use of in vivo testing. However, human skin equivalents are already being used by pharmaceutical and cosmetic companies, mainly in the preliminary phases for product development to screen the potential toxicity of unknown materials [7]. Most of the reconstructed human epidermal models commercialized today include EpiSkin®, SkinEthic®, and Epiderm®, while living skin equivalents are commercialized by GraftSkin®, EpidermFT®, and Phenion®. These human skin analogs are the standards used in the testing of irritation, corrosion and sensitization tests used in most companies and they may serve as models for the development of new skin analogs, for instance for animal skin analogs, since they are approved and regulated and follow the OECD test guidelines for chemicals [2,8–10].

Indeed, animal skin models are needed for the development of pharmaceutical and cosmetic products for veterinary use, yet there are no canine skin equivalents commercially available. Although, some methodologies were described to maintain canine skin biopsies in culture [11,12], to isolate dermal fibroblasts and keratinocytes and to obtain a 3D multilayer epidermis using natural canine skin explants [4,13,14], there is a lack of an easily standardizable methodology using culture-developed canine epidermis to test the efficacy and the biosafety of veterinary products. Thus, the aim of this study was to develop a canine epidermis analog based on CPEK (canine progenitor epithelial keratinocytes) expansion and differentiation in culture at the air–liquid surface. Tissue patches developed by this strategy are highly reproduceable and useful as a skin surrogate for veterinary product development.

2. Materials and Methods

For the development of the cEpiderm, epidermal keratinocyte progenitor cells supplied by CELLnTEC (Bern, Switzerland) were used. The 3D keratinocyte Starter Protocol [15] was followed with some adaptations. Briefly, 2×10^6 cells were seeded in uncoated inserts (VWR Internacional, Carnaxide, Portugal), in DMEM (Sigma-Aldrich, Darmstadt, Germany) supplemented with 10% fetal bovine serum (FBS; Biowest, Nuaille, France), 5 ng/mL insulin (Sigma-Aldrich, Darmstadt, Germany) and 10 ng/mL epithelial growth factor (EGF; Sino Biological, Eschborn, Germany). The culture medium inside and outside the inserts was changed every 2–3 days until confluence was reached, which typically occurred 7–10 days after seeding. After the confirmation of the confluence, the culture media was changed, inside and outside the insert, for differentiation medium (CnT-PR-3D, CELLnTEC). The following day, the inserts were air-lifted to promote differentiation, and changing of the medium under the insert was carried out every 2–3 days until an epidermis-like tissue was obtained (see Figure 1).

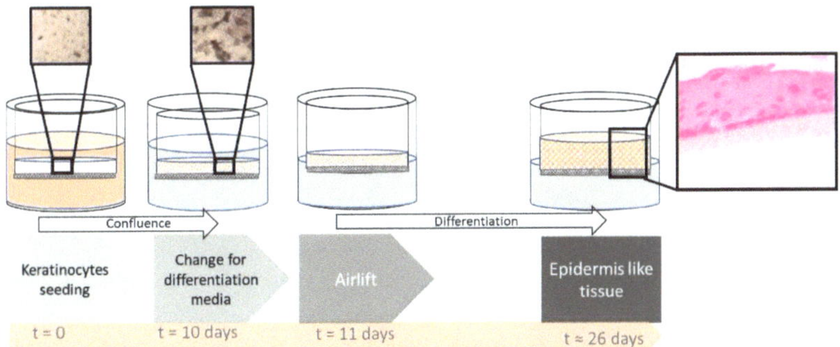

Figure 1. Schematic representation of the keratinocyte differentiation process. Microscopic photographs of the cell culture in the upper boxes showing the cell density (at seeding time and after 10-day culture) and a histological image of the differentiated tissue in the side box (stained with eosin-hematoxylin).

For histological analysis, tissue patches were collected and fixed in 10% neutral-buffered formalin (VWR Internacional, Carnaxide, Portugal) and processed for examination by standard light microscopy techniques (Olympus Iberica, Llobregat, Spain). Paraffin (Panreac AppliChem, Darmstadt, Germany) sections were cut at 3 μm and stained with hematoxylin-eosin (Harries Hematoxylin modified solution for clinical diagnosis, Panreac AppliChem, Darmstadt, Germany; Eosin Y, Merk, Darmstadt, Germany) [16].

To assess cell barrier impermeability, 1% Triton-X100 (Sigma-Aladrich, Darmstadt, Germany) was added on top of the epidermal tissue and incubated over 4 h at 37 °C [14], and subsequently cell viability was measured with CCK-8 (Sigma Aldrich, Darmstadt, Germany) [17].

Corrosion and irritation tests were performed using a validated methodology for human skin models [8,9]. Briefly, stimuli were performed using 5% SDS (Himedia, Einhausen, Germany) and glacial acetic acid (Fisher Chemical, Porto Salvo, Portugal) for the irritation and the corrosion tests, respectively. PBS (phosphate buffer saline, in mM: Na_2HPO_4 10.0; KH_2PO_4 1.8; NaCl 137.0; KCl 2.7; pH 7.4) was used as control. Tissue patches were incubated with the stimuli for 42 min at room temperature for the irritation test, and 3 min and 60 min at 37 °C for the corrosion tests. After incubation, the inserts were washed with PBS to remove any traces of the stimuli, and CCK-8 was added to assess cell viability according to the supplier's instructions [8–10,17].

3. Results and Discussion

An air-lift keratinocyte's histotypical culture was produced and morphological, histological and functional assessments were performed to evaluate its canine epidermal-like (cEpiderm) properties.

3.1. Morphological and Histological Characterization of the cEpiderm

The tissue patches obtained were constituted by a multilayer of canine keratinocytes, 3–4 cell layers thick, generating a stratified epidermal-like tissue, confirmed by histological analysis (Figure 2).

The epidermal-like tissue (Figure 2A) exhibited cohesive keratinocytes (single layer of keratinized cells) comparable to those observed in the squamous epithelial cells of skin epidermis (Figure 2B). The keratinocytes in the histotypical model are cuboidal or slightly flattened cells that are attached to adjacent cells via desmosomes, which is also a feature of canine epidermis.

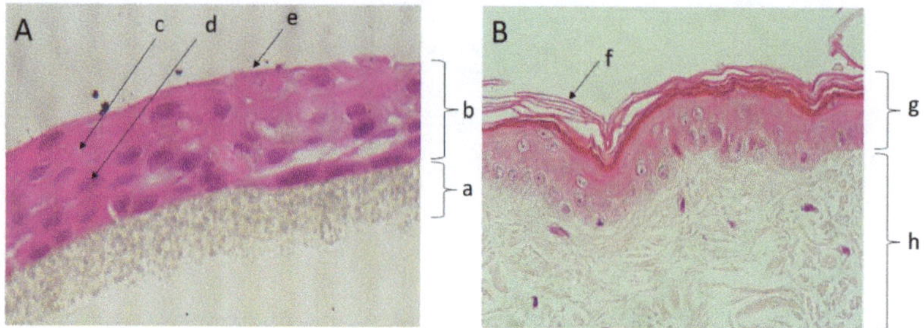

Figure 2. Micrographs of a section of the canine epidermis-like culture (**A**) and a canine skin biopsy (**B**), both routinely processed and stained with hematoxylin and eosin (400× amplification). Legend: (a) insert porous membrane; (b) keratinocyte layers of the histotypical culture; (c) cell junctions; (d) keratinocyte nucleus; (e) stratum corneum-like layer; (f) stratum corneum; (g) epidermis; (h) dermis.

Our results are in a good agreement with those observed by Yagihara and collaborators (2011), whereby using inserts coated with type-I collagen allowed for the observation that CPEK cells cultured at an air–liquid interface became stratified and formed characteristic epidermis layers [18].

3.2. Functional Characterization of the cEpiderm

Exposing the cEpiderm apical side to 0.1% Triton X-100 evoked a 45% reduction in cell viability (Figure 3A), revealing a good impermeability performance, an important feature of a functional epidermal-like tissue [9].

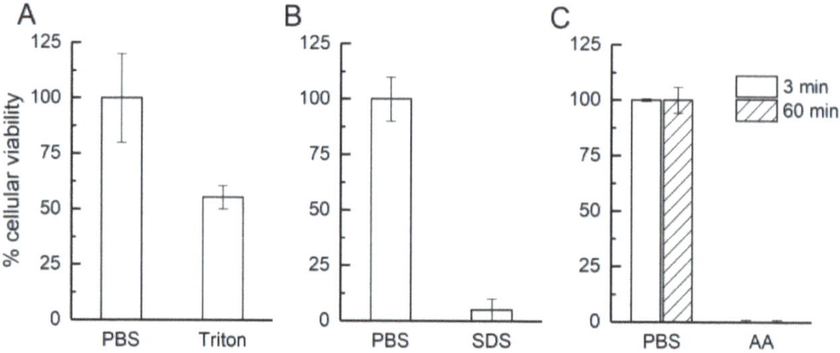

Figure 3. Cellular viability accessed in permeability (**A**), irritation (**B**), and corrosion (**C**) tests. PBS was used as control; Triton: 0.1% Triton X-100; SDS: 5% SDS; AA: acetic acid (glacial).

For functional irritation response, the cEpiderm slices were exposed on the apical side to 5% SDS. The negatively charged detergent induced a loss of cell viability of 95% compared to controls (Figure 3B). Histological evaluation of the biopsies revealed that SDS acted by disrupting the structure of cEpiderm patches, causing detachments of the keratinocytes (Figure 4, upper panel).

For this test, a known irritant (SDS) was used and the cEpiderm response was similar to the reference tables, corresponding to a loss of cell viability above 50% [8,10].

After a corrosion insult with acetic acid (glacial), the cell viability declined to 0% after 3 min of exposure, remaining at 0% after 60 min (Figure 3C). Histological evaluation has shown that the cell patches remain attached but show a progressive disruption of the tissue, characterized by a loss of integrity at the tissue surface (see arrows in Figure 4, lower panel).

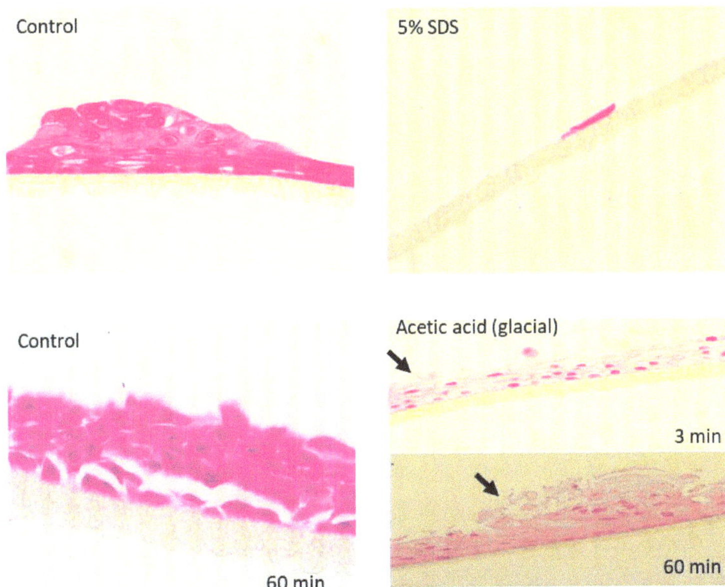

Figure 4. Micrographs of sections of the cEpiderm untreated (Control) and upon irritation (5% SDS) or corrosion (acetic acid) insult. Arrows point to tissue disruption.

Acetic acid (glacial) is a known subcategory 1A corrosive chemical which causes an acute massive loss of cell viability (>95%) [9,10], consistent with the response of the cEpiderm model.

The cEpiderm responded as expected to corrosion, irritation and permeability assessment tests, and in accordance with the observations in equivalent human in vitro tests [8]. These are good indicators of cEpiderm's functional barrier and response to stressors [10], suggesting that it is suitable to be used in biosafety and biocompatibility screening studies. This canine epidermis model presents the properties to be used for other common tests in drug and cosmetics development, but also for other stimuli assessment (e.g., phototoxicity, response to UV-lights), permeability and/or skin sensitivity (evaluating the epithelial immune activation). Considering the above, the canine epidermis model described in this paper has the potential to greatly contribute to the reduction of animal testing in the development of veterinary products. On the other hand, the cEpiderm skin model lacks a circulation system, thus having limited applicability to evaluate angiogenesis, a process involved in wound healing in vivo [5].

One limitation of this work was the absence of reference tables for the irritation and corrosion in vitro tests for canine skin or epidermis, such as those already available for human skin models [19]. In the future, guidelines for corrosion and irritation tests appliable for in vitro canine epidermis should be defined, so the cEpiderm may be validated as a canine skin surrogate.

4. Conclusions

A stratified epidermal-like tissue, cEpiderm, which responds to irritation and corrosion insults, according to the OECD guidelines for human skin models, was successfully developed using commercially available keratinocytes (CPEK) grown in air-lift uncoated inserts. The cEpiderm has a high repeatability potential and is suitable for large-scale use.

The use of species-specific histotypical models for epidermis may contribute to more accurate safety results and to a significative reduction in animal testing in the development of veterinary pharmaceuticals and cosmetic products. The high standardization potential of this epidermis model makes it suitable for the future development of guidelines for animal-free canine skin testing.

Author Contributions: Conceptualization, A.R.C. and C.M.A.; methodology, J.N. and M.M.; validation, S.B. and L.M.; formal analysis, J.N. and M.M.; investigation, B.U., L.F., J.N. and M.M.; resources, A.J.B., C.F. and L.M.; data curation, A.R.C. and C.M.A.; writing—original draft preparation, A.R.C., C.M.A., J.N. and M.M.; writing—review and editing, A.J.B., A.R.C. and C.M.A., S.B.; supervision, A.R.C. and C.M.A.; project administration, A.J.B., C.F. and L.M.; funding acquisition, A.C.C. and A.J.B. All authors have read and agreed to the published version of the manuscript.

Funding: This research was funded by Alentejo 2020 Programme, grant number ALT20-03-247-FEDER-033578, and co-financed by FEDER Funds through the Operational Programme for Competitiveness Factors—COMPETE 2020 and by National Funds through FCT—Foundation for Science and Technology under the Project UIDB/50006/2020, UIDP/5006/2020, UIDB/04683/2020 and UIDP/04683/2020.

Conflicts of Interest: The authors declare no conflict of interest.

References

1. Netzlaff, F.; Kaca, M.; Bock, U.; Haltner-Ukomadu, E.; Meiers, P.; Lehr, C.-M.; Schaefer, U.F. Permeability of the reconstructed human epidermis model Episkin in comparison to various human skin preparations. *Eur. J. Pharm. Biopharm.* **2007**, *66*, 127–134. [CrossRef] [PubMed]

2. MacNeil, S. Progress and opportunities for tissue-engineered skin. *Nature* **2007**, *445*, 874–880. [CrossRef] [PubMed]

3. Supp, D.M.; Boyce, S.T. Engineered skin substitutes: Practices and potentials. *Clin. Dermatol.* **2005**, *23*, 403–412. [CrossRef] [PubMed]

4. Souci, L.; Denesvre, C. 3D skin models in domestic animals. *Vet. Res.* **2021**, *52*, 21. [CrossRef] [PubMed]

5. Rittié, L. Cellular mechanisms of skin repair in humans and other mammals. *J. Cell Commun. Signal.* **2016**, *10*, 103–120. [CrossRef]

6. Suhail, S.; Sardashti, N.; Jaiswal, D.; Rudraiah, S.; Misra, M.; Kumbar, S.G. Engineered Skin Tissue Equivalents for Product Evaluation and Therapeutic Applications. *Biotechnol. J.* **2019**, *14*, e1900022. [CrossRef]

7. Martínez-Santamaría, L.; Guerrero-Aspizua, S.; Del Río, M. Skin bioengineering: Preclinical and clinical applications. *Actas Dermosifiliogr.* **2012**, *103*, 5–11. [CrossRef] [PubMed]

8. Skin Irritation Protocol SkinEthic RHE, (Validated) [Internet]. Lyon, FRANCE: Episkin. Available online: https://www.episkin.com/skin-irritation (accessed on 2 May 2022).

9. Skin Corrosion Protocol Episkin (Validated) [Internet]. Lyon, FRANCE: Episkin. Available online: https://www.episkin.com/skin-corrosion (accessed on 2 May 2022).

10. Netzlaff, F.; Lehr, C.M.; Wertz, P.W.; Schaefer, U.F. The human epidermis models EpiSkin®, SkinEthic® and EpiDerm®: An evaluation of morphology and their suitability for testing phototoxicity, irritancy, corrosivity, and substance transport. *Eur. J. Pharm. Biopharm.* **2005**, *60*, 167–178. [CrossRef] [PubMed]

11. Abramo, F.; Pirone, A.; Lenzi, C.; Vannozzi, I.; Della Valle, M.F.; Miragliotta, V. Establishment of a 2-week canine skin organ culture model and its pharmacological modulation by epidermal growth factor and dexamethasone. *Ann. Anat.* **2016**, *207*, 109–117. [CrossRef]

12. Bauhammer, I.; Sacha, M.; Haltner, E. Establishment of an in vitro model of cultured viable human, porcine and canine skin and comparison of different media supplements. *PeerJ* **2019**, *7*, e7811. [CrossRef]

13. Serra, M.; Brazís, P.; Puigdemont, A.; Fondevila, D.; Romano, V.; Torre, C.; Ferrer, L. Development and characterization of a canine skin equivalent. *Exp. Dermatol.* **2007**, *16*, 135–142. [CrossRef] [PubMed]

14. Cerrato, S.; Ramió-Lluch, L.; Fondevila, D.; Rodes, D.; Brazis, P.; Puigdemont, A. Effects of Essential Oils and Polyunsaturated Fatty Acids on Canine Skin Equivalents: Skin Lipid Assessment and Morphological Evaluation. *J. Vet. Med.* **2013**, *2013*, 231526. [CrossRef] [PubMed]

15. 3D Keratinocyte Starter Kit (without Primary Cells) [Internet]. BERN, SWITZERLAND: CELLnTEC. Available online: https://cellntec.com/wp-content/uploads/pdf/3D_Starter_Kit_K.pdf (accessed on 2 May 2022).

16. Histological Processing and Staining [Internet]. BERN, SWITZERLAND. Available online: https://cellntec.com/wp-content/uploads/pdf/RoutineHistology_Staining.pdf (accessed on 2 May 2022).

17. Cell Counting Kit-8 [Internet]. Darmstadt, Germany: Sigma-Aldrich. Available online: https://www.sigmaaldrich.com/deepweb/assets/sigmaaldrich/product/documents/381/017/96992dat.pdf (accessed on 2 May 2022).

18. Yagihara, H.; Okumura, T.; Shiomi, E.; Shinozaki, N.; Kuroki, S.; Sasaki, Y.; Ito, K.; Ono, K.; Washizu, T.; Bonkobara, M. Reconstruction of stratum corneum in organotypically cultured canine keratinocyte-derived CPEK cells. *Vet. Res. Commun.* **2011**, *35*, 433–437. [CrossRef] [PubMed]

19. OECD. Test No. 431: In vitro skin corrosion: Reconstructed human epidermis (RHE) test method. In *OECD Guidelines for the Testing of Chemicals, Section 4*; OECD: Paris, France, 2019.

Article

Nitric Oxide Production from Nitrite plus Ascorbate during Ischemia upon Hippocampal Glutamate NMDA Receptor Stimulation

Carla Nunes [1,2,*] and **João Laranjinha** [1,2]

[1]   Center for Neuroscience and Cell Biology, University of Coimbra, 3004-504 Coimbra, Portugal
[2]   Faculty of Pharmacy, University of Coimbra, 3000-548 Coimbra, Portugal
*   Correspondence: cmnunes@cnc.uc.pt

Abstract: Nitric oxide ($^\bullet$NO), a diffusible free radical, is an intercellular messenger, playing a crucial role in several key brain physiological processes, including in neurovascular coupling (NVC). In the brain, glutamatergic activation of the neuronal nitric oxide synthase (nNOS) enzyme constitutes its main synthesis pathway. However, when oxygen (O_2) supply is compromised, such as in stroke, ischemia, and aging, such $^\bullet$NO production pathway may be seriously impaired. In this context, evidence suggests that, as already observed in the gastric compartment, the reduction of nitrite by dietary compounds (such as ascorbate and polyphenols) or by specific enzymes may occur in the brain, constituting an important rescuing or complementary mechanism of $^\bullet$NO production. Here, using microsensors selective for $^\bullet$NO, we show that nitrite enhanced the $^\bullet$NO production in a concentration-dependent manner and in the presence of ascorbate evoked by N-methyl-D-aspartate (NMDA) and glutamate stimulation of rat hippocampal slices. Additionally, nitrite potentiated the $^\bullet$NO production induced by oxygen-glucose deprivation (OGD). Overall, these observations support the notion of a redox interaction of ascorbate with nitrite yielding $^\bullet$NO upon neuronal glutamatergic activation and given the critical role of NO as the direct mediator of neurovascular coupling may represents a key physiological mechanism by which $^\bullet$NO production for cerebral blood flow (CBF) responses to neuronal activation is sustained under hypoxic/acidic conditions in the brain.

Keywords: nitric oxide; nitrite; ascorbate; hippocampal slices

Citation: Nunes, C.; Laranjinha, J. Nitric Oxide Production from Nitrite plus Ascorbate during Ischemia upon Hippocampal Glutamate NMDA Receptor Stimulation. *BioChem* **2023**, *3*, 78–90. https://doi.org/10.3390/biochem3020006

Academic Editor: Nidhi Jalan-Sakrikar

Received: 6 February 2023
Revised: 13 March 2023
Accepted: 6 April 2023
Published: 3 May 2023

1. Introduction

Although the human brain represents only 2% of body weight, it consumes approximately 20% of the oxygen (O_2) and 25% of glucose utilized by the body and receives nearly 15% of the cardiac output [1]. These high energetic and metabolic demands of the human brain, associated with its very limited reserve capacity, imply a continuous and tightly regulated cerebral blood flow (CBF) to assure a proper supply of metabolic substrates, namely O_2 and glucose, as well as the clearance of metabolic waste byproducts [2]. In this context, it is important to note that even a small reduction in CBF has a deleterious impact on brain function [3,4]. As such, a mechanism termed neurovascular coupling (NVC), which involves a tight network communication between all the cells that comprise the Neurovascular Unit (neurons, glia and cerebrovascular cells), ensures a fine temporal and regional regulation of the CBF according to neuronal activity to fulfill the metabolic needs [5]. Recently, strong evidence has emerged that the highly diffusible free radical, nitric oxide ($^\bullet$NO), plays a pivotal role in the NVC [6–8]. In the Central Nervous System (CNS), $^\bullet$NO is also involved in other brain physiological processes, such as learning and memory formation, synaptic plasticity, mitochondrial respiration and modulation of neurotransmitter release [9–13].

The canonical pathway for $^\bullet$NO synthesis in mammalian cells is carried out by a family of enzymes, the nitric oxide synthases (NOS) [14]. These enzymes catalyze the oxidation of L-arginine to L-citrulline and $^\bullet$NO, using NADPH and oxygen (O_2) as co-substrates [14]. There are three major isoforms of NOS: neuronal NOS (nNOS or NOS 1), endothelial NOS (eNOS or NOS 3) and inducible NOS (iNOS or NOS 2), all of them present in the CNS. Both nNOS and eNOS are constitutively expressed and activated by Ca^{+2}/calmodulin-dependent signaling, producing nanomolar concentrations of $^\bullet$NO for seconds or minutes [15]. iNOS is induced in glial cells following immunological or inflammatory stimulation, and its activity is Ca^{2+}-independent, producing high concentrations of $^\bullet$NO for hours or days [15].

$^\bullet$NO signaling in the brain is intimately associated with glutamatergic neurotransmission [16]. In neurons, the activation of N-methyl-D-aspartate receptor (NMDAr) by glutamate binding leads to the influx of Ca^{+2} that, upon binding to calmodulin, activates nNOS, which is physically coupled to NMDAr by the scaffolding protein postsynaptic density-95 (PSD-95) [17]. In addition, at the capillary level, glutamate may activate the NMDAr in the endothelial cells and thus leads to eNOS activation.

As NOS requires O_2 to work properly, in situations where a decrease in O_2 supply occurs, such as stroke, ischemia, and aging, the activity of constitutive NOS is compromised, and thereby, the enzymatic $^\bullet$NO production can significantly decrease, leading to an impairment of NVC and other central $^\bullet$NO functions. However, in recent years, some studies point to nitrite as a key bioprecursor of $^\bullet$NO in the brain, particularly under acidic and hypoxic conditions [18–21], as it was already verified in the gastric compartment [22,23]. In this context, it was demonstrated that acute nitrite enhanced basal CBF in a rat model [24] and also recovered NVC to its original magnitude in a rat model of somatosensory stimulation under conditions of NOS inhibition [25].

Notably, nitrite concentration in the body may be enhanced through the diet, namely by ingesting green leafy vegetables, which are rich in nitrate [26]. Nitrate is quickly absorbed across the upper gastrointestinal tract [27,28]. Although much of the nitrate may be excreted in the urine, up to 25% is taken up by salivary glands, where it is concentrated and released into the oral cavity. Here, nitrate is reduced to nitrite by oral commensal bacteria [29]. In the stomach, a nitrite may be reduced to $^\bullet$NO [30,31]. Components of the diet, such as ascorbate and polyphenols, may enhance nitrite reduction [22,23]. The remaining nitrite is absorbed in the small intestine, and through circulation, it can reach several organs [18], including the brain, where increased nitrite levels are observed in the cerebrospinal fluid [32]. A growing body of evidence supports that the reduction of nitrite to $^\bullet$NO may be a possibility in the brain, namely by powerful reducing agents such as ascorbate and polyphenols [20,33], especially under low oxygen tension. In accordance, Presley and colleagues showed that a high-nitrate diet increases regional brain perfusion in older subjects in specific brain areas [34].

In this work, using microsensors with a high degree of selectivity to $^\bullet$NO, we explore the non-NOS production of $^\bullet$NO via nitrite reduction in rat hippocampal slices stimulated with NMDA and glutamate and also subjected to oxygen-glucose deprivation (OGD).

2. Materials and Methods

2.1. Materials and Reagents

NMDA was purchased from Hello Bio (Bristol, UK). Ortho-phenylenediamine (O-PD), Nafion$^\circledR$, ascorbic acid, L-glutamic acid and sodium nitrite were from Sigma-Aldrich. A Carbox gas mixture (O_2/CO_2) was obtained from Linde, Lisbon, Portugal. All solutions were prepared in MilliQ water, bi-deionized water ultrapure with resistivity higher than 18 MΩ.cm (Millipore Corporation, Burlington, MA, USA).

For hippocampal slice assays, we used artificial cerebrospinal fluid (aCSF) with the following composition (in mM): 124 NaCl, 2 KCl, 25 NaHCO3, 1.25 NaH_2PO_4, 1.5 $CaCl_2$, 0.1 $MgCl_2$ and 10 D-glucose. In order to improve the viability of the slices, a modified aCSF was used for slice dissection and recovery. The composition of this aCSF was (in mM): 124 NaCl, 2 KCl, 25 NaHCO3, 1.25 NaH_2PO_4, 1.5 $CaCl_2$, 1 $MgCl_2$, 1 reduced glutathione

(GSH), 0.2 ascorbate and 10 D-glucose. Both aCSFs were continuously bubbled with Carbox for oxygenation and pH buffering (pH 7.4).

2.2. Electrochemical Instrumentation

All recordings were performed using a Compactstat Potentiostat (Ivium, Eindhoven, The Netherlands). For $^\bullet$NO and O_2 hippocampal recordings, a two-electrode circuit was used with the microsensor as the working electrode and an Ag/AgCl (3M NaCl) as a reference electrode. The working electrode was held at a constant potential of +0.7 V or −0.8 V vs. Ag/AgCl for $^\bullet$NO and O_2 measurements, respectively.

2.3. Carbon Fiber Microelectrode Fabrication and Surface Modification

Carbon fiber microelectrodes (CFM) were essentially fabricated as previously described [35]. Briefly, a single carbon fiber ($\varnothing$ = 30 µm, Textron, Lowell, MA, USA) was inserted into a borosilicate glass capillary (Science Products GmbH, Hofheim, Germany). The capillaries were pulled on a vertical puller (Harvard Apparatus Ltd., Cambourne, UK), and then the protruding carbon fibers were cut under a microscope to obtain a tip length of 150–200 µm. The electrical contact was provided by injecting a small portion of a conductive silver paint (RS Pro, Corby, UK) into the capillary, followed by the insertion of a copper wire with the outer insulation previously removed from the extremities. The microelectrodes were tested for their general recording properties in phosphate-buffered saline (PBS) medium by fast cyclic voltammetry (FCV) at a 200 V/s scan rate between −1.0 and 1.0 V vs. Ag/AgCl for 30 s (EI400 potentiostat, Ensman Instruments, Bloomigton, USA). The microelectrodes that passed in FCV evaluation (stable background current and sharp transients at reversal potentials) were modified in a two-step protocol with Nafion® and O-PD to improve their analytical properties for $^\bullet$NO measurements. First, the tips of the CFMs were immersed in Nafion® solution (5% in aliphatic alcohols) for ten seconds, followed by drying at 180 °C for 5 min. High temperatures seem to enhance the adherence of Nafion® film to the carbon fiber surface [36]. After cooling, the process was repeated. On the day of the use, the CFMs were coated with O-PD by electropolymerization at a constant potential of +0.7 V vs. Ag/AgCl for 30 min.

2.4. Carbon Fiber Microelectrodes Calibration

Each $^\bullet$NO microelectrode was evaluated for $^\bullet$NO sensitivity and selectivity towards nitrite and ascorbate by constant voltage amperometry at +0.7 V vs. Ag/AgCl.

The CFMs for O_2 measurement were evaluated regarding their sensitivity towards O_2. After deoxygenation of PBS (40 mL) by bubbling nitrogen in a sealed vessel, several additions of 200 µL of an O_2-saturated solution, which corresponds to 6.22 µM of O_2, were performed. The sensitivity of the CFMs for O_2 was determined based on the slope of the calibration curve.

2.5. Rat Hippocampal Slices

All animal procedures used in this study were performed in accordance with the European Union Council Directive for the Care and Use of Laboratory Animals, 2010/63/EU, and were approved by the local ethics committee (ORBEA) and the Portuguese Directorate-General for Food and Veterinary. Male Wistar rats with ages between 6–7 weeks were decapitated under deep anesthesia. The brain was quickly removed and placed in ice-cold modified aCSF saturated with Carbox. The hippocampi were dissected, and transverse slices with a thickness of 400 µm were obtained with a McIlwain tissue chopper (Campden Instruments, London, UK). The separated slices were transferred into a prerecording chamber containing modified aCSF at room temperature and continuously bubbled with Carbox. Slices were maintained under these conditions at least 1 h prior to use, allowing for good tissue recovery.

2.6. Nitric Oxide and Oxygen Recording

Individual slices were placed in a recording chamber (BSC-BU with BSC-ZT top; Harvard Apparatus) and perfused with normal aCSF at 32 °C (temperature controller model TC-202A; Harvard Apparatus) bubbled with Carbox. The electrodes were inserted into the pyramidal cell layer of the CA1 subregion of the rat hippocampal slice, a region that is easy to identify and with a high expression of glutamatergic receptors and nNOS. The hippocampal slices were stimulated with NMDA (100 μM) or glutamate (5 mM) added to the perfusion aCSF for 2 min.

For OGD studies, slices placed in the recording chamber were perfused with aCSF without glucose and saturated with a gas mixture of 95% of nitrogen gas (N2) and 5% CO_2 for 10 min. Then, reoxygenation was performed by replacing the perfused medium with normal aCSF saturated with Carbox.

To measure the pH variation during OGD, a pH microelectrode ($\varnothing$ = 100 μm, Unisense, Denmark) and an O_2 microelectrode were inserted in an antiparallel orientation into the pyramidal cell layer of the CA1 subregion of the rat hippocampal slice.

2.7. Data Analysis

Using the OriginPro 7.5 software (OriginLab Corporation, Northampton, MA, USA), the recorded •NO signals were individually analyzed in terms of (1) the •NO peak amplitude of the signal; (2) the signal area, calculated as the time integral of the •NO signal; and (3) the •NO signal width as the time in seconds from the stimulation point to return to basal levels.

All statistical analyses were performed using GraphPad Prism 5 software (GraphPad Software, San Diego, CA, USA). Data are presented as mean ± SEM. Statistical analysis of the data was performed using one-way analysis of variance (ANOVA) followed by the post hoc Bonferroni's multiple comparison test. Differences were considered significant at $p < 0.05$.

3. Results

3.1. Nitric Oxide Concentration Dynamics in Rat Hippocampus Slices Evoked by NMDA Stimulation of Neuronal Activity: The Modulatory Role of Nitrite

The hippocampal slices were stimulated with NMDA, the synthetic and specific agonist of the NMDA receptor (NMDAr), present in the perfusion medium for a 2 min period in the absence (control) and presence of three different concentrations of nitrite (100 μM, 500 μM and 2 mM). Ascorbate (500 μM) was always present in the aCSF during all the perfusions. This concentration of ascorbate was selected on the basis of our previous studies showing the functional coupling of ascorbate and nitric oxide dynamics in the rat hippocampus upon stimulation of NMDAr that induces the release of ascorbate from stimulated neurons, achieving a local extracellular concentration of about 500 μM [33,37]. Stimulation of the NMDAr causes a local oxygen tension drop (data not shown) and induces a biphasic •NO production in the CA1 subregion (Figure 1A). As observed in Figure 1A, the first component of the signal temporally coincides with the period of neuronal stimulation. Notably, in the presence of nitrite and in a concentration-dependent manner, both components of the •NO signal exhibit a greater amplitude (Figure 1A,B) and a greater area (Figure 1A,C). Regarding the •NO signal width, no significant differences are observed between the control condition and the presence of nitrite (Figure 1A,D).

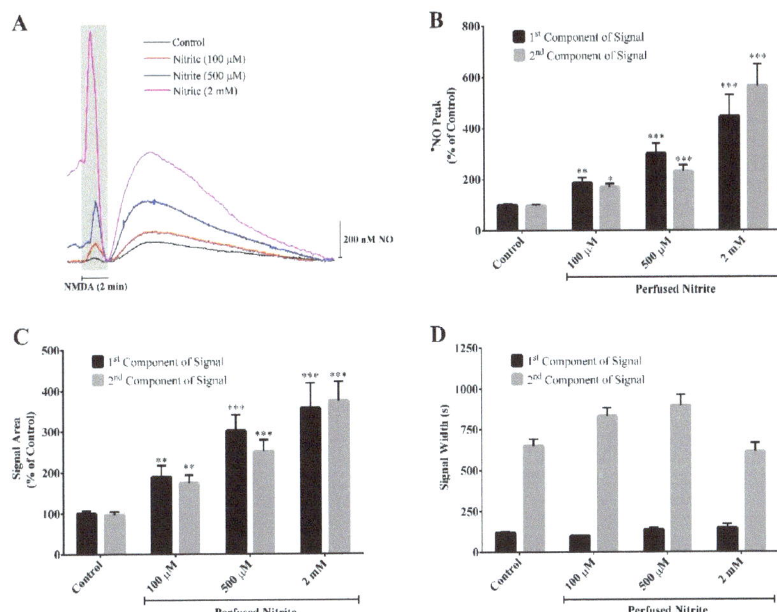

Figure 1. Nitrite enhances the production of •NO evoked by perfusion of NMDA (100 μM) for 2 min (shade area) in the presence of ascorbate (500 μM). Recordings were performed in the CA1 subregion of rat hippocampal slices. Ascorbate is present in the perfusion medium. (**A**) Representative amperometric recording of the •NO production in the absence (control) and presence of different nitrite concentrations (100 μM, 500 μM and 2 mM). (**B**) Amplitude, (**C**) area and (**D**) duration of the biphasic •NO signal. The first and the second components of the signal correspond to the first and the second peak of the biphasic •NO signal, respectively. Each bar represents the mean ± SEM. Statistical significance: * $p < 0.05$, ** $p < 0.01$ and *** $p < 0.001$ as compared with the control.

3.2. Nitric Oxide Concentration Dynamics in Rat Hippocampus Slices upon Stimulation of Neuronal Activity by Glutamate

Hippocampal slices perfused with aCSF containing ascorbate 500 μM were also stimulated with glutamate, the endogenous agonist of glutamate receptors, in the absence (control) and presence of nitrite (100 μM). It was found that, in the presence of nitrite, the signals corresponding to the production of •NO have a greater amplitude (Figure 2A,B), a greater area (Figure 2A,C) and a longer duration (Figure 2A,D). The •NO concentration dynamics are distinct from that obtained with the artificial NMDA receptor agonist (NMDA), notably the absence of a biphasic signal, likely reflecting more complex mechanisms of activation that translate into different kinetics of •NO production by the neuronal •NO isoform associated with the NMDA receptor.

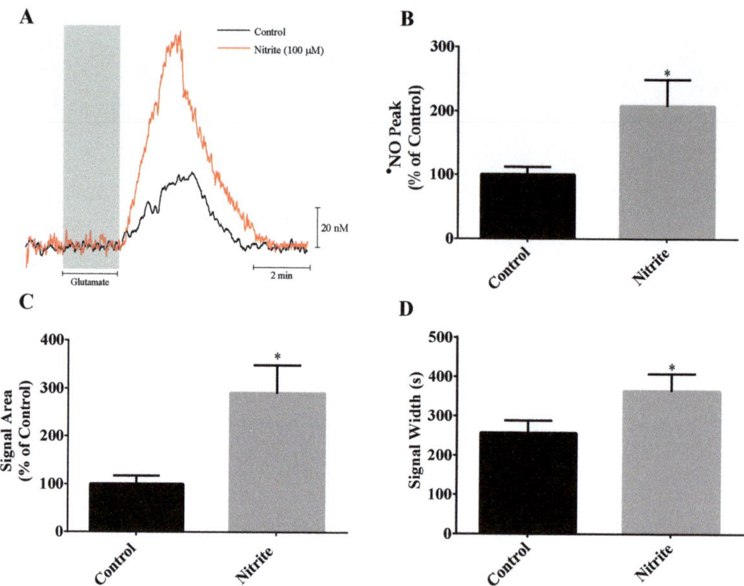

Figure 2. Nitrite enhances the production of •NO evoked by perfusion of glutamate (5 mM) for 2 min in the presence of ascorbate (500 μM). Recordings were performed in the CA1 subregion of rat hippocampal slices. Ascorbate is present in the perfusion medium. (**A**) Representative amperometric recording of the •NO production in the absence (control) and presence of nitrite (100 μM). (**B**) Amplitude, (**C**) area and (**D**) duration of the •NO concentration dynamics. Each bar represents the mean ± SEM. Statistical significance: * $p < 0.05$ as compared with the control.

3.3. Evaluation of pH and Nitric Oxide Production in Rat Hippocampus Slices in Transient Oxygen-Glucose Deprivation (OGD)

The production of •NO via the reduction of nitrite may be particularly relevant under hypoxic and acidotic conditions, a situation in which the enzymatic •NO synthesis might be impaired as •NO synthases use O_2 as a substrate. Therefore, using both a pH microelectrode and a selective CFM for O_2 measurement, pH and O_2 variations were simultaneously monitored in rat hippocampal slices subjected to hypoxia for 10 min, followed by reoxygenation. It is observed that, upon OGD conditions, the pH value gradually decreases from 7.4 during the hypoxia period, reaching a minimum of approximately 7.17 (Figure 3A). About 1.5 min after the start of reoxygenation, the pH gradually increases until it reaches the normal value of 7.4 (Figure 3A).

Also, •NO and O_2 were simultaneously measured under OGD conditions in the absence (control) and presence of nitrite (100 μM, 500 μM and 2 mM) followed by reoxygenation. In the absence of added nitrite, we observe a signal detected by the •NO microsensor during OGD and a further residual signal following NMDA (100 μM) stimulation for 2 min (Figure 3B). Both signals increase as a function of the nitrite concentration in the perfusion media, from 100 μM (Figure 3C), 500 μM (Figure 3D) to 2 mM (Figure 3E). In order to verify if this signal corresponded to •NO production, the experiments were repeated under an applied potential of 0.4 V to the CFM, a potential at which oxidation of •NO does not occur. Under these conditions, the signal almost disappears (Figure 3F). Figure 4 shows the quantification of the variations shown in Figure 3. It is of note that the signals obtained upon stimulation with NMDA (Figure 3B–E) under OGD conditions are smaller as compared with •NO production observed in hippocampal slices not subjected to OGD (Figure 1).

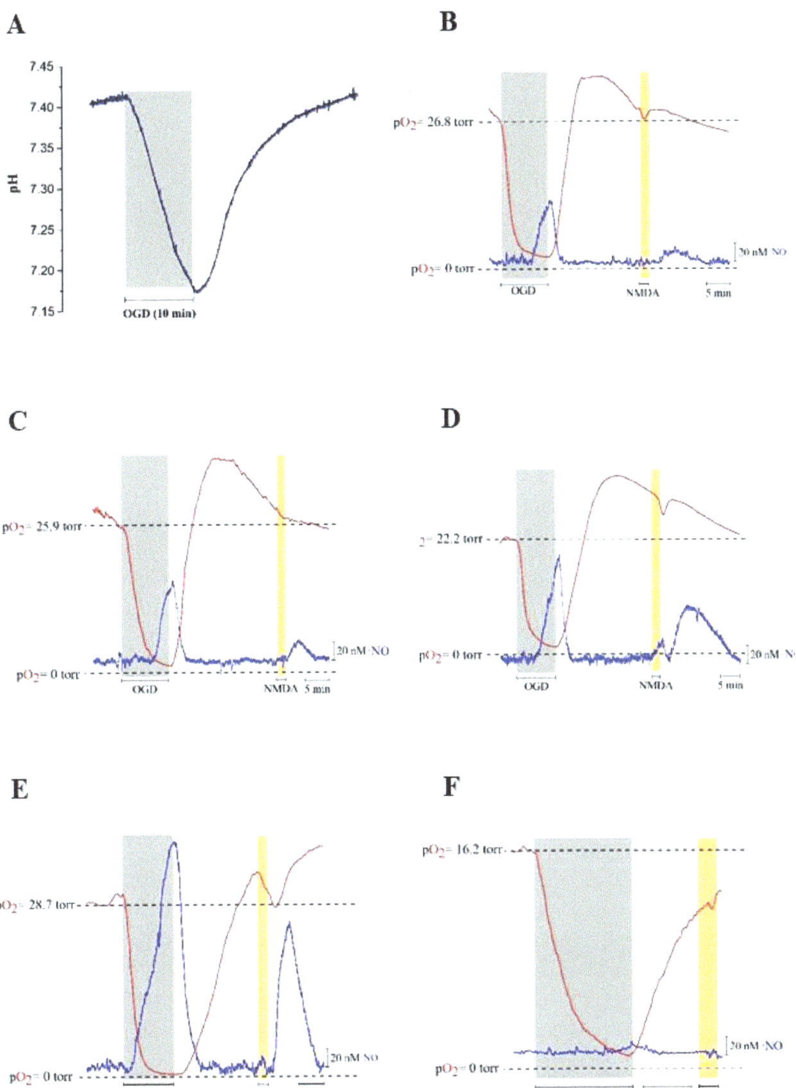

Figure 3. Nitrite enhances •NO production during OGD for 10 min and by perfusion of NMDA (100 µM) for 2 min after reoxygenation. Recordings were performed in the CA1 subregion of rat hippocampal slices. Ascorbate (500 µM) is present in the perfusion medium. (**A**) Representative recording of pH inside of the rat hippocampal slice during OGD for 10 min. Simultaneous recording of •NO (blue line) and O_2 (red line) in the (**B**) absence of nitrite (control), presence of (**C**) 100 µM nitrite, (**D**) 500 µM nitrite, (**E**) 2 mM nitrite and (**F**) absence of nitrite at 0.4 V.

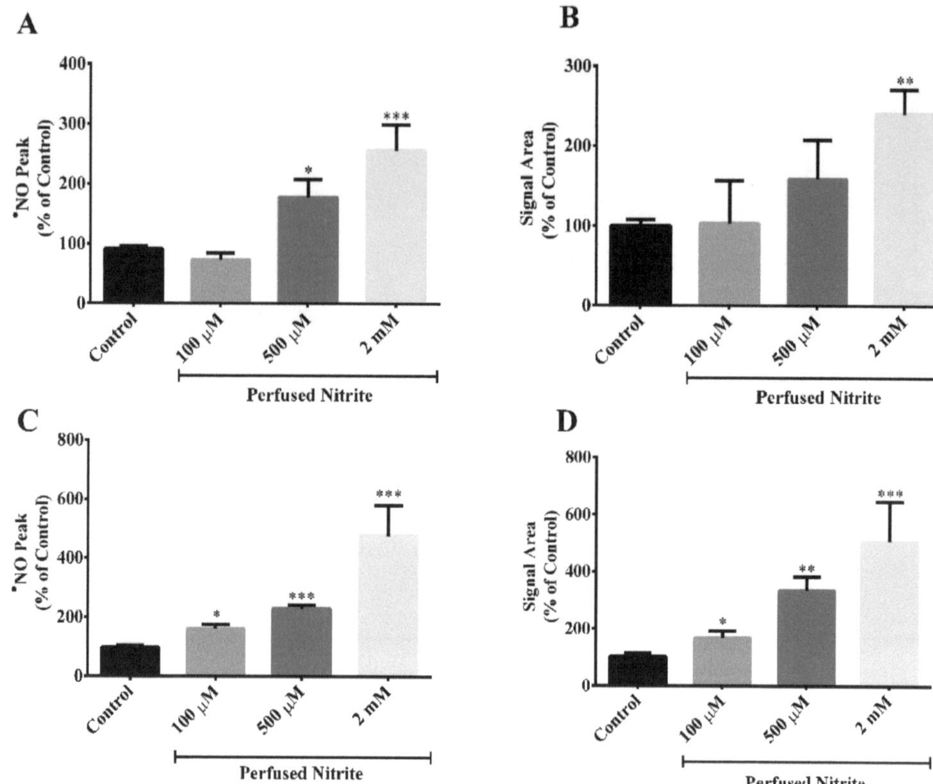

Figure 4. Quantification of $^\bullet$NO production during OGD in the presence of nitrite and by perfusion of NMDA (100 µM) for 2 min after reoxygenation. Recordings were performed in the CA1 subregion of rat hippocampal slices. Ascorbate (500 µM) is present in the perfusion medium. (**A**) Amplitude and (**B**) area of the $^\bullet$NO signal evoked by OGD. (**C**) Amplitude and (**D**) area of the $^\bullet$NO signal evoked by perfusion of NMDA (100 µM) for 2 min after reoxygenation. Each bar represents the mean $\pm$ SEM. Statistical significance: * $p < 0.05$, ** $p < 0.01$ and *** $p < 0.001$ as compared with the control.

4. Discussion

While the synthesis of $^\bullet$NO by NOS is O_2 dependent, the nitrate–nitrite–$^\bullet$NO pathway is gradually activated as tensions of O_2 drop, seeming to play a key role in vasodilation [38], modulation of mitochondrial respiration [39], hypoxia [40] and also in the tissue protection in ischemia-reperfusion [41].

Although the one-electron reduction of nitrite to $^\bullet$NO by dietary reductants is well established in the gastric compartment [22,23], this pathway of $^\bullet$NO production has been poorly explored in the brain environment. In this context, it is worth mentioning that ascorbate is a powerful reducing agent that exists in abundance in the brain, particularly in the cerebral cortex and hippocampus, where it may reach concentrations around 10 mM in neurons [42–44]. Moreover, there is evidence for functional coupling of ascorbate and $^\bullet$NO dynamics in the rat hippocampus upon stimulation of NMDAr [33], triggering ascorbate release from neuronal cells to the extracellular fluid during glutamatergic activity, probably by the glutamate–ascorbate heteroexchange mechanism [45–47], with ascorbate concentrations of up to 500 µM reported in the cerebrospinal fluid [42–44] and the extracellular neuronal environment [37].

Under hypoxic or ischemic conditions, such as in stroke, ischemia and aging, the NOS activity is impaired, the local pH drops and the reduction of nitrite to $^\bullet$NO may constitute an important rescuing or alternative mechanism to the classical NOS pathway under these conditions [19,40]. The hippocampus seems to be the cerebral area most susceptible to hypoxic injury [48,49]. Importantly, several years ago, Millar proposed the reduction of nitrite to $^{\bullet}$NO in the extracellular space by ascorbate released from neurons under increased activity as a regulatory mechanism of vascular oxygen delivery according to the local metabolic needs of neurons [20]. Indeed, as neuronal stimulation causes local O_2 drop to very low tensions [50,51] (creating a "hypoxia-like" transient status), the reduction of nitrite to $^{\bullet}$NO by extracellular ascorbate concomitantly released from neurons is known to be favored under these conditions [52]. Although some studies showed that, under hypoxia, the pH might drop to values near 6.4, causing severe brain acidosis [53–55]. Here, we observed that OGD for 10 min induced a mild brain acidosis in the CA1 region of rat hippocampal slices, as the pH fell from 7.4 to 7.17. This may be justified by differences regarding the model of this study, the age of the animals and the OGD time.

In agreement with Millar's hypothesis, here we show that during conditions simulating ischemia-reperfusion conditions (transient OGD) and in the absence of neuronal nitric oxide synthase (nNOS) activation, a transient increase of $^{\bullet}$NO is observed in a nitrite concentration-dependent manner and in ascorbate-containing perfusion media (Figure 3). Residual nitrite contained in the slices might account for the $^{\bullet}$NO signal under control conditions (Figure 3B). In agreement with these results, it has been reported that there is a protective action of nitrite during ischemia and ischemia-reperfusion in several organs, including in the brain [56–58], and this protection is attributed to its reduction in $^{\bullet}$NO.

These observations strongly support nitrite as the source of $^{\bullet}$NO under conditions in which nNOS might be inoperative due to diminished O_2 bioavailability. Moreover, it is of note that, following OGD and under normal conditions, the stimulation with NMDA induces $^{\bullet}$NO transients that, although expected, are quantitatively smaller as compared with the OGD for each of the recordings (Figure 3B–E).

Expectedly, different kinetic traces are obtained with NMDA and glutamate that likely reflect the complexity of the pathways leading to NMDAr activation by glutamate as compared with NMDA, but, as in the case of NMDA stimulation, the glutamate-dependent $^{\bullet}$NO transients are dependent on the concentration of nitrite in the perfusion medium.

It should be mentioned that controls in the absence of ascorbate in the perfusion media in all the experiments induced little quantitative effects of NO concentration dynamics. As mentioned before, our previous studies supported a functional coupling of ascorbate and $^{\bullet}$NO dynamics in the rat hippocampus upon stimulation of NMDAr that induces the release of ascorbate from stimulated neurons [33,37]. In fact, considering the millimolar range of ascorbate in neurons, the concentration of ascorbate released into the extracellular medium upon glutamatergic stimulation could be extremely high and, thereby, further increasing ascorbate concentration via its addition to the perfusion medium did not translate into an increase of nitrite reduction. Furthermore, some specific enzymes, such as hemoglobin [39,59], xanthine oxidoreductase [60,61], complexes of the mitochondrial electron transport chain [62–64] and even the NOS [65], seem to acquire a nitrite reductase activity, particularly at low O_2 tension, and may be involved in the $^{\bullet}$NO production from nitrite. Thus, we wanted to make sure the presence of enough ascorbate to reduce nitrite in all the several conditions, including those of OGD where ascorbate release from neurons is not coupled to NMDAr activation.

The concentrations of nitrite used in this study (100 μM, 500 μM and 2 mM) might be considered exceedingly high concentrations. However, in our model of study, hippocampal slices were perfused with the aCSF that contains nitrite, with a nitrite gradient along the thickness of the slice being expected. Because the recordings are performed in the core of the slice, we chose these concentrations to ensure that an adequate concentration of nitrite reached the core of the slice, local to where the microsensors were inserted. Moreover, these concentrations permit to clearly evaluate whether the effect of nitrite was concentration-dependent.

5. Conclusions

Overall, this work strongly suggests that nitrite is a precursor of •NO in the brain in a way that is functionally dependent on glutamate NMDAr activation via the NMDAr–nNOS pathway but also under conditions in which nNOS might be impaired, notably hypoxia. Given the critical role of •NO as the direct mediator of neurovascular coupling (NVC [7], a process critical for brain structure and function [66]), the mechanism of •NO production from nitrite via its reduction to •NO by ascorbate may represent a key physiological mechanism that participates in the regulation of brain homeostasis by improving the NVC and CBF and could acquire a particular relevance in situations where NVC is compromised, such as ischemia, neurodegeneration and aging. Therefore, the nitrite–•NO pathway may open avenues to the development of new and innovative therapies to improve the impaired NVC observed in several conditions, including neurodegenerative diseases, aging and diabetes [55,66,67].

Author Contributions: C.N.: conceptualization, formal analysis, investigation, and writing original draft preparation. J.L.: conceptualization, supervision, funding acquisition, and writing—review. All authors have read and agreed to the published version of the manuscript.

Funding: This work was financed by the European Regional Development Fund (ERDF) through the COMPETE 2020 (Operational Programme for Competitiveness and Internationalization) and by the Portuguese National Funds via FCT—Fundação para a Ciência e Tecnologia—under projects POCI-01-0145-FEDER-029099, 2022.05454.PTDC, UIDB/04539/2020, UIDP/04539/2020 and LA/P/0058/2020.

Institutional Review Board Statement: The animal study protocol was approved by the local ethics committee (ORBEA 146_2016/31102016) and the Portuguese Directorate-General for Food and Veterinary.

Informed Consent Statement: Not applicable.

Data Availability Statement: The data presented in this study are available upon request from the corresponding author.

Acknowledgments: Authors are thankful to Cátia F. Lourenço for their help in the preparation of the microelectrodes.

Conflicts of Interest: The authors declare no conflict of interest.

References

1. Magistretti, P.J.; Pellerin, L.; Rothman, D.L.; Shulman, R.G. Energy on demand. *Science* **1999**, *283*, 496–497. [CrossRef]
2. Jackman, K.; Iadecola, C. Neurovascular regulation in the ischemic brain. *Antioxid. Redox Signal.* **2015**, *22*, 149–160. [CrossRef] [PubMed]
3. Drake, C.T.; Iadecola, C. The role of neuronal signaling in controlling cerebral blood flow. *Brain Lang.* **2007**, *102*, 141–152. [CrossRef] [PubMed]
4. Hossmann, K.A. Viability thresholds and the penumbra of focal ischemia. *Ann. Neurol.* **1994**, *36*, 557–565. [CrossRef] [PubMed]
5. Iadecola, C. The Neurovascular Unit Coming of Age: A Journey through Neurovascular Coupling in Health and Disease. *Neuron* **2017**, *96*, 17–42. [CrossRef]
6. Lourenco, C.F.; Laranjinha, J. Nitric Oxide Pathways in Neurovascular Coupling Under Normal and Stress Conditions in the Brain: Strategies to Rescue Aberrant Coupling and Improve Cerebral Blood Flow. *Front. Physiol.* **2021**, *12*, 729201. [CrossRef]
7. Lourenco, C.F.; Santos, R.M.; Barbosa, R.M.; Cadenas, E.; Radi, R.; Laranjinha, J. Neurovascular coupling in hippocampus is mediated via diffusion by neuronal-derived nitric oxide. *Free Radic. Biol. Med.* **2014**, *73*, 421–429. [CrossRef]
8. Hosford, P.S.; Gourine, A.V. What is the key mediator of the neurovascular coupling response? *Neurosci. Biobehav. Rev.* **2019**, *96*, 174–181. [CrossRef]
9. Calabrese, V.; Mancuso, C.; Calvani, M.; Rizzarelli, E.; Butterfield, D.A.; Stella, A.M. Nitric oxide in the central nervous system: Neuroprotection versus neurotoxicity. *Nat. Rev. Neurosci.* **2007**, *8*, 766–775. [CrossRef]
10. Laranjinha, J.; Nunes, C.; Ledo, A.; Lourenco, C.; Rocha, B.; Barbosa, R.M. The Peculiar Facets of Nitric Oxide as a Cellular Messenger: From Disease-Associated Signaling to the Regulation of Brain Bioenergetics and Neurovascular Coupling. *Neurochem. Res.* **2021**, *46*, 64–76. [CrossRef]
11. Laranjinha, J.; Santos, R.M.; Lourenco, C.F.; Ledo, A.; Barbosa, R.M. Nitric oxide signaling in the brain: Translation of dynamics into respiration control and neurovascular coupling. *Ann. N. Y. Acad. Sci.* **2012**, *1259*, 10–18. [CrossRef] [PubMed]
12. Picon-Pages, P.; Garcia-Buendia, J.; Munoz, F.J. Functions and dysfunctions of nitric oxide in brain. *Biochim. Et Biophys. Acta Mol. Basis Dis.* **2019**, *1865*, 1949–1967. [CrossRef]

13. Ledo, A.; Lourenco, C.F.; Cadenas, E.; Barbosa, R.M.; Laranjinha, J. The bioactivity of neuronal-derived nitric oxide in aging and neurodegeneration: Switching signaling to degeneration. *Free Radic. Biol. Med.* **2021**, *162*, 500–513. [CrossRef] [PubMed]

14. Alderton, W.K.; Cooper, C.E.; Knowles, R.G. Nitric oxide synthases: Structure, function and inhibition. *Biochem. J.* **2001**, *357*, 593–615. [CrossRef]

15. Moncada, S.; Palmer, R.M.; Higgs, E.A. Nitric oxide: Physiology, pathophysiology, and pharmacology. *Pharm. Rev.* **1991**, *43*, 109–142.

16. Ledo, A.; Barbosa, R.M.; Gerhardt, G.A.; Cadenas, E.; Laranjinha, J. Concentration dynamics of nitric oxide in rat hippocampal subregions evoked by stimulation of the NMDA glutamate receptor. *Proc. Natl. Acad. Sci. USA* **2005**, *102*, 17483–17488. [CrossRef] [PubMed]

17. Christopherson, K.S.; Hillier, B.J.; Lim, W.A.; Bredt, D.S. PSD-95 assembles a ternary complex with the N-methyl-D-aspartic acid receptor and a bivalent neuronal NO synthase PDZ domain. *J. Biol. Chem.* **1999**, *274*, 27467–27473. [CrossRef]

18. Lundberg, J.O.; Weitzberg, E.; Gladwin, M.T. The nitrate-nitrite-nitric oxide pathway in physiology and therapeutics. *Nat. Rev. Drug Discov.* **2008**, *7*, 156–167. [CrossRef]

19. Shiva, S. Nitrite: A Physiological Store of Nitric Oxide and Modulator of Mitochondrial Function. *Redox Biol.* **2013**, *1*, 40–44. [CrossRef]

20. Millar, J. The nitric oxide/ascorbate cycle: How neurones may control their own oxygen supply. *Med. Hypotheses* **1995**, *45*, 21–26. [CrossRef]

21. Kapil, V.; Khambata, R.S.; Jones, D.A.; Rathod, K.; Primus, C.; Massimo, G.; Fukuto, J.M.; Ahluwalia, A. The Noncanonical Pathway for In Vivo Nitric Oxide Generation: The Nitrate-Nitrite-Nitric Oxide Pathway. *Pharm. Rev.* **2020**, *72*, 692–766. [CrossRef]

22. Gago, B.; Lundberg, J.O.; Barbosa, R.M.; Laranjinha, J. Red wine-dependent reduction of nitrite to nitric oxide in the stomach. *Free Radic. Biol. Med.* **2007**, *43*, 1233–1242. [CrossRef] [PubMed]

23. Rocha, B.S.; Gago, B.; Barbosa, R.M.; Laranjinha, J. Dietary polyphenols generate nitric oxide from nitrite in the stomach and induce smooth muscle relaxation. *Toxicology* **2009**, *265*, 41–48. [CrossRef]

24. Rifkind, J.M.; Nagababu, E.; Barbiro-Michaely, E.; Ramasamy, S.; Pluta, R.M.; Mayevsky, A. Nitrite infusion increases cerebral blood flow and decreases mean arterial blood pressure in rats: A role for red cell NO. *Nitric Oxide* **2007**, *16*, 448–456. [CrossRef] [PubMed]

25. Piknova, B.; Kocharyan, A.; Schechter, A.N.; Silva, A.C. The role of nitrite in neurovascular coupling. *Brain Res.* **2011**, *1407*, 62–68. [CrossRef]

26. Brkic, D.; Bosnir, J.; Bevardi, M.; Boskovic, A.G.; Milos, S.; Lasic, D.; Krivohlavek, A.; Racz, A.; Cuic, A.M.; Trstenjak, N.U. Nitrate in Leafy Green Vegetables and Estimated Intake. *Afr. J. Tradit. Complement. Altern. Med.* **2017**, *14*, 31–41. [CrossRef] [PubMed]

27. Miyoshi, M.; Kasahara, E.; Park, A.M.; Hiramoto, K.; Minamiyama, Y.; Takemura, S.; Sato, E.F.; Inoue, M. Dietary nitrate inhibits stress-induced gastric mucosal injury in the rat. *Free Radic. Res.* **2003**, *37*, 85–90. [CrossRef]

28. Witter, J.P.; Balish, E.; Gatley, S.J. Distribution of nitrogen-13 from labeled nitrate and nitrite in germfree and conventional-flora rats. *Appl. Env. Microbiol.* **1979**, *38*, 870–878. [CrossRef]

29. Duncan, C.; Dougall, H.; Johnston, P.; Green, S.; Brogan, R.; Leifert, C.; Smith, L.; Golden, M.; Benjamin, N. Chemical generation of nitric oxide in the mouth from the enterosalivary circulation of dietary nitrate. *Nat. Med.* **1995**, *1*, 546–551. [CrossRef]

30. Benjamin, N.; O'Driscoll, F.; Dougall, H.; Duncan, C.; Smith, L.; Golden, M.; McKenzie, H. Stomach NO synthesis. *Nature* **1994**, *368*, 502. [CrossRef]

31. Lundberg, J.O.; Weitzberg, E.; Lundberg, J.M.; Alving, K. Intragastric nitric oxide production in humans: Measurements in expelled air. *Gut* **1994**, *35*, 1543–1546. [CrossRef] [PubMed]

32. Pluta, R.M.; Dejam, A.; Grimes, G.; Gladwin, M.T.; Oldfield, E.H. Nitrite infusions to prevent delayed cerebral vasospasm in a primate model of subarachnoid hemorrhage. *Jama* **2005**, *293*, 1477–1484. [CrossRef] [PubMed]

33. Ferreira, N.R.; Lourenco, C.F.; Barbosa, R.M.; Laranjinha, J. Coupling of ascorbate and nitric oxide dynamics in vivo in the rat hippocampus upon glutamatergic neuronal stimulation: A novel functional interplay. *Brain Res. Bull.* **2015**, *114*, 13–19. [CrossRef]

34. Presley, T.D.; Morgan, A.R.; Bechtold, E.; Clodfelter, W.; Dove, R.W.; Jennings, J.M.; Kraft, R.A.; King, S.B.; Laurienti, P.J.; Rejeski, W.J.; et al. Acute effect of a high nitrate diet on brain perfusion in older adults. *Nitric Oxide* **2011**, *24*, 34–42. [CrossRef]

35. Santos, R.M.; Lourenco, C.F.; Piedade, A.P.; Andrews, R.; Pomerleau, F.; Huettl, P.; Gerhardt, G.A.; Laranjinha, J.; Barbosa, R.M. A comparative study of carbon fiber-based microelectrodes for the measurement of nitric oxide in brain tissue. *Biosens. Bioelectron.* **2008**, *24*, 704–709. [CrossRef] [PubMed]

36. Gerhardt, G.A.; Hoffman, A.F. Effects of recording media composition on the responses of Nafion-coated carbon fiber microelectrodes measured using high-speed chronoamperometry. *J. Neurosci. Methods* **2001**, *109*, 13–21. [CrossRef]

37. Ferreira, N.R.; Santos, R.M.; Laranjinha, J.; Barbosa, R.M. Real Time In Vivo Measurement of Ascorbate in the Brain Using Carbon Nanotube-Modified Microelectrodes. *Electroanalysis* **2013**, *25*, 1757–1763. [CrossRef]

38. Cosby, K.; Partovi, K.S.; Crawford, J.H.; Patel, R.P.; Reiter, C.D.; Martyr, S.; Yang, B.K.; Waclawiw, M.A.; Zalos, G.; Xu, X.; et al. Nitrite reduction to nitric oxide by deoxyhemoglobin vasodilates the human circulation. *Nat. Med.* **2003**, *9*, 1498–1505. [CrossRef]

39. Shiva, S.; Huang, Z.; Grubina, R.; Sun, J.; Ringwood, L.A.; MacArthur, P.H.; Xu, X.; Murphy, E.; Darley-Usmar, V.M.; Gladwin, M.T. Deoxymyoglobin is a nitrite reductase that generates nitric oxide and regulates mitochondrial respiration. *Circ. Res.* **2007**, *100*, 654–661. [CrossRef]

40. van Faassen, E.E.; Bahrami, S.; Feelisch, M.; Hogg, N.; Kelm, M.; Kim-Shapiro, D.B.; Kozlov, A.V.; Li, H.; Lundberg, J.O.; Mason, R.; et al. Nitrite as regulator of hypoxic signaling in mammalian physiology. *Med. Res. Rev.* **2009**, *29*, 683–741. [CrossRef]

41. Shiva, S.; Sack, M.N.; Greer, J.J.; Duranski, M.; Ringwood, L.A.; Burwell, L.; Wang, X.; MacArthur, P.H.; Shoja, A.; Raghavachari, N.; et al. Nitrite augments tolerance to ischemia/reperfusion injury via the modulation of mitochondrial electron transfer. *J. Exp. Med.* **2007**, *204*, 2089–2102. [CrossRef]

42. Mefford, I.N.; Oke, A.F.; Adams, R.N. Regional distribution of ascorbate in human brain. *Brain Res.* **1981**, *212*, 223–226. [CrossRef]

43. Milby, K.; Oke, A.; Adams, R.N. Detailed mapping of ascorbate distribution in rat brain. *Neurosci. Lett.* **1982**, *28*, 15–20. [CrossRef] [PubMed]

44. Rice, M.E.; Russo-Menna, I. Differential compartmentalization of brain ascorbate and glutathione between neurons and glia. *Neuroscience* **1998**, *82*, 1213–1223. [CrossRef] [PubMed]

45. Miele, M.; Boutelle, M.G.; Fillenz, M. The physiologically induced release of ascorbate in rat brain is dependent on impulse traffic, calcium influx and glutamate uptake. *Neuroscience* **1994**, *62*, 87–91. [CrossRef]

46. Sandstrom, M.I.; Rebec, G.V. Extracellular ascorbate modulates glutamate dynamics: Role of behavioral activation. *BMC Neurosci.* **2007**, *8*, 32. [CrossRef]

47. Cammack, J.; Ghasemzadeh, B.; Adams, R.N. The pharmacological profile of glutamate-evoked ascorbic acid efflux measured by in vivo electrochemistry. *Brain Res.* **1991**, *565*, 17–22. [CrossRef] [PubMed]

48. Shaw, K.; Bell, L.; Boyd, K.; Grijseels, D.M.; Clarke, D.; Bonnar, O.; Crombag, H.S.; Hall, C.N. Neurovascular coupling and oxygenation are decreased in hippocampus compared to neocortex because of microvascular differences. *Nat. Commun.* **2021**, *12*, 3190. [CrossRef]

49. Zhang, H.; Roman, R.J.; Fan, F. Hippocampus is more susceptible to hypoxic injury: Has the Rosetta Stone of regional variation in neurovascular coupling been deciphered? *GeroScience* **2022**, *44*, 127–130. [CrossRef]

50. Ledo, A.; Barbosa, R.; Cadenas, E.; Laranjinha, J. Dynamic and interacting profiles of *NO and O2 in rat hippocampal slices. *Free Radic. Biol. Med.* **2010**, *48*, 1044–1050. [CrossRef]

51. Park, S.S.; Hong, M.; Song, C.K.; Jhon, G.J.; Lee, Y.; Suh, M. Real-time in vivo simultaneous measurements of nitric oxide and oxygen using an amperometric dual microsensor. *Anal. Chem.* **2010**, *82*, 7618–7624. [CrossRef] [PubMed]

52. Lundberg, J.O.; Weitzberg, E. Nitrite reduction to nitric oxide in the vasculature. *Am. J. Physiol. Heart Circ. Physiol.* **2008**, *295*, H477–H478. [CrossRef] [PubMed]

53. Funahashi, T.; Floyd, R.A.; Carney, J.M. Age effect on brain pH during ischemia/reperfusion and pH influence on peroxidation. *Neurobiol. Aging* **1994**, *15*, 161–167. [CrossRef]

54. Nemoto, E.M.; Frinak, S. Brain tissue pH after global brain ischemia and barbiturate loading in rats. *Stroke* **1981**, *12*, 77–82. [CrossRef]

55. von Hanwehr, R.; Smith, M.L.; Siesjo, B.K. Extra- and intracellular pH during near-complete forebrain ischemia in the rat. *J. Neurochem.* **1986**, *46*, 331–339. [CrossRef]

56. Jung, K.H.; Chu, K.; Ko, S.Y.; Lee, S.T.; Sinn, D.I.; Park, D.K.; Kim, J.M.; Song, E.C.; Kim, M.; Roh, J.K. Early intravenous infusion of sodium nitrite protects brain against in vivo ischemia-reperfusion injury. *Stroke* **2006**, *37*, 2744–2750. [CrossRef] [PubMed]

57. Dias, C.; Lourenco, C.F.; Laranjinha, J.; Ledo, A. Modulation of oxidative neurometabolism in ischemia/reperfusion by nitrite. *Free Radic. Biol. Med.* **2022**, *193*, 779–786. [CrossRef]

58. Luettich, A.; Franko, E.; Spronk, D.B.; Lamb, C.; Corkill, R.; Patel, J.; Ezra, M.; Pattinson, K.T.S. Beneficial Effect of Sodium Nitrite on EEG Ischaemic Markers in Patients with Subarachnoid Haemorrhage. *Transl. Stroke Res.* **2022**, *13*, 265–275. [CrossRef] [PubMed]

59. Gladwin, M.T.; Kim-Shapiro, D.B. The functional nitrite reductase activity of the heme-globins. *Blood* **2008**, *112*, 2636–2647. [CrossRef]

60. Cantu-Medellin, N.; Kelley, E.E. Xanthine oxidoreductase-catalyzed reduction of nitrite to nitric oxide: Insights regarding where, when and how. *Nitric Oxide* **2013**, *34*, 19–26. [CrossRef]

61. Godber, B.L.; Doel, J.J.; Sapkota, G.P.; Blake, D.R.; Stevens, C.R.; Eisenthal, R.; Harrison, R. Reduction of nitrite to nitric oxide catalyzed by xanthine oxidoreductase. *J. Biol. Chem.* **2000**, *275*, 7757–7763. [CrossRef] [PubMed]

62. Castello, P.R.; David, P.S.; McClure, T.; Crook, Z.; Poyton, R.O. Mitochondrial cytochrome oxidase produces nitric oxide under hypoxic conditions: Implications for oxygen sensing and hypoxic signaling in eukaryotes. *Cell Metab.* **2006**, *3*, 277–287. [CrossRef] [PubMed]

63. Kozlov, A.V.; Staniek, K.; Nohl, H. Nitrite reductase activity is a novel function of mammalian mitochondria. *FEBS Lett.* **1999**, *454*, 127–130. [CrossRef] [PubMed]

64. Nohl, H.; Staniek, K.; Sobhian, B.; Bahrami, S.; Redl, H.; Kozlov, A.V. Mitochondria recycle nitrite back to the bioregulator nitric monoxide. *Acta Biochim. Pol.* **2000**, *47*, 913–921. [CrossRef]

65. Vanin, A.F.; Bevers, L.M.; Slama-Schwok, A.; van Faassen, E.E. Nitric oxide synthase reduces nitrite to NO under anoxia. *Cell. Mol. Life Sci.* **2007**, *64*, 96–103. [CrossRef]

66. Lourenco, C.F.; Ledo, A.; Barbosa, R.M.; Laranjinha, J. Neurovascular uncoupling in the triple transgenic model of Alzheimer's disease: Impaired cerebral blood flow response to neuronal-derived nitric oxide signaling. *Exp. Neurol.* **2017**, *291*, 36–43. [CrossRef]
67. Goncalves, J.S.; Seica, R.M.; Laranjinha, J.; Lourenco, C.F. Impairment of neurovascular coupling in the hippocampus due to decreased nitric oxide bioavailability supports early cognitive dysfunction in type 2 diabetic rats. *Free Radic. Biol. Med.* **2022**, *193*, 669–675. [CrossRef]

 **BioChem**

Communication

UALGORITMO, a New Instrument of the University of Algarve for Scientific Outreach

José Bragança [1,2,3,*], Sónia Figueiredo [4], Carla Alexandra Rego [5], Filomena dos Reis Conceição [6] and Saúl Neves de Jesus [7,8]

[1] Faculdade de Medicina e Ciências Biomédicas (FMCB), Universidade do Algarve, Campus de Gambelas, 8005-139 Faro, Portugal
[2] Algarve Biomedical Center Research Institute, ABC-RI, 8005-139 Faro, Portugal
[3] Champalimaud Research Program, Champalimaud Center for the Unknown, 1400-038 Lisbon, Portugal
[4] School José Belchior Viegas, Sítio da Calçada, 8150-021 São Brás de Alportel, Portugal; sonia.figueiredo@aejbv.pt
[5] Schools Pinheiro e Rosa, 8005-546 Faro, Portugal; carego@aeprosa.pt
[6] High School Tomás Cabreira, 8000-075 Faro, Portugal; filomena.conceicao@agr-tc.pt
[7] Faculdade de Ciências Humanas e Sociais, Universidade do Algarve, 8005-139 Faro, Portugal; snjesus@ualg.pt
[8] Centro de Investigação em Turismo, Sustentabilidade e Bem-estar (CinTurs), Universidade do Algarve, Campus de Gambelas, 8005-139 Faro, Portugal
* Correspondence: jebraganca@ualg.pt

Abstract: Researchers at Universities generate and convey the knowledge acquired through communications in specialized (inter)national journals and congresses. An effort to share the scientific achievements with the general public is extremely important. For this purpose, we have launched the UALGORITMO, a journal freely accessible online, written in lay Portuguese language by Researchers of the University of the Algarve, to summarize recent communications published in peer reviewed journals. After submission, the manuscripts are revised by High Schools Students of the Algarve, under the guidance of a schoolteacher, for further simplification of the language and general improvement of the manuscript and figures. The revised manuscripts by the authors are edited and published, with an acknowledgment and a presentation of the reviewers at the end of each article. To maximize the outreach, the articles include a summarized biography of the authors, and links to their research centers and teaching courses. We believe that the UALGORITMO is a valuable instrument to promote scientific literacy and culture amongst all communities.

Keywords: outreach; layman; high school education; public engagement; science communication; scientific literacy

Citation: Bragança, J.; Figueiredo, S.; Rego, C.A.; dos Reis Conceição, F.; Neves de Jesus, S. UALGORITMO, a New Instrument of the University of Algarve for Scientific Outreach. *BioChem* **2022**, *2*, 93–103. https://doi.org/10.3390/biochem2010007

Academic Editors: Manuel Aureliano, M. Leonor Cancela, Célia M. Antunes and Ana Rodrigues Costa

Received: 20 December 2021
Accepted: 24 February 2022
Published: 3 March 2022

Publisher's Note: MDPI stays neutral with regard to jurisdictional claims in published maps and institutional affiliations.

1. Introduction

Universities have as their principal mission to generate and share knowledge, but these institutions also have to contribute to the education of all surrounding communities. Thus, the scientific dissemination and communication to the general public contributes to the latter vocation of the Universities, as it presents the value of the scientific production in simpler terms, as well as the potential benefits of these research activities for each individual in the society and for the society as a whole. In academia, most of the research activities are funded by governmental, charity, and/or philanthropic organizations, which strongly value the importance of the communication of the scientific results originated with their support, not only in renowned and specialized meetings and peer reviewed journals, but also through the direct engagement in public and lay outreach actions [1]. Scientific outreach is also very important to convey scientific literacy to the public and may be considered as an educational tool that can be used across students' communities to enrich their curricular training, and to encourage them to engage in future scientific careers [2–4].

In fact, science communication is not only essential for sharing exciting new scientific achievements to the public, but also has an important role in providing knowledge, understanding and appreciation of science, both to the general public and non-scientist decision-makers, for taking better consensual decisions about individual and collective lives, health, and happiness [5]. Moreover, with the exponential scientific progress happening continuously in all research and engineering domains, for science communication to be efficient, scientists and other science conveyers have to communicate the right message obtained from the most reliable and accurate sources of information to the right audience with the most appropriate and accessible tools [3–6]. Bearing this idea in mind, in recent years, numerous scientific communication journals have been directed to a variety of readers, such as full-time researchers, to help them to keep up to date with all scientific publication of interest, science conveyers, teachers, and students of all ages, as well as families.

The strategies developed by scientists for public outreach are numerous, such as presentations of the research activities during open days at the institutions, lectures to lay public, interventions and interviews in traditional news media or novel social platform media, publications of books or videos for lay public, and through both institutional and personal websites. The strategies may also vary according to the scientists' age and status. Indeed, studies across several countries have indicated that older scientists are generally more incline to participate in outreach activities than younger colleagues [7]. Perhaps this situation is due to the fact that older scientists are likely to hold leadership positions with more time to spare for events that are beyond the scope of the research activities, while younger scientists are more focused and involved in working on their research projects and career progression. Moreover, older scientists prefer the presentations in traditional news media and books, while younger scientists would rather participate in events in direct contact with the public [1]. In any case, the most commonly reported obstacles by scientists for engaging in outreach events with the public are the lack of time necessary for these activities, the absence of specific funding, and no institutional incentive for these activities [8]. In addition, scientists often perceive themselves as poorly trained, and with little confidence in their communication skills, to efficiently participate in outreach activities and to exhilarate the public [3,7]. Interestingly, the most consistent factors to encourage scientists to take part in outreach activities are the conviction that they will enjoy the activity, be efficient in the transmission of their message and information to the public, and of course, have the time to engage in these activities [7].

The University of Algarve has a long-lasting track record of interactions and engagement with the public, in particular to primary and secondary schools of the Algarve through diversified initiatives. Indeed, the University of Algarve organizes free open days and site visits to the University, and summer classes for students and teachers of primary and secondary schools. Lectures and other activities jointly coordinated with schools, and adapted to the requests and needs of the school students and teachers, are also organized either at the University campuses or at the schools. The University of Algarve and its research centers have also obtained funding for the development of an Itinerant Mobile Molecular Genetics Laboratory (Lab-IT), which proposes hands-on practical laboratory sessions at secondary schools using basic molecular biology techniques. The Lab-It also promotes molecular genetics knowledge and disseminates applications of molecular biology in biomedical, pharmaceutical, forensic, biotechnological, and environmental sciences, amongst other domains. In addition, the University of Algarve collaborates closely with the Ciência Viva Center of the Algarve (http://www.ccvalg.pt/, accessed on 15 December 2021), which was the first interactive Center of the Portuguese Ciência Viva Centers network, dedicated precisely to the dissemination of scientific and technological knowledge to lay public, in particular to the youth community. The Ciência Viva Center of the Algarve has a permanent exposition on local sea resources and wildlife, environment and green energy, and activities related to biology, physics, and chemistry, which are often closely involved with projects developed at the University of Algarve, with the participation of its researchers. Finally, researchers of the University of Algarve are frequently invited to share their new findings or to comment on novel scientific advances in mainstream and specialized national media. Here, we describe the UALGORITMO, a journal freely accessible online created as a novel outreach instrument of the University of Algarve.

2. The UALGORITMO, a Collaboration between the University and Secondary High Schools

The UALGORITMO was developed as an instrument to disseminate all recent research, technological, and art activities of the University of Algarve, using an approach previously unexplored that is complementary to the other outreach strategies already implemented at the University. The UALGORITMO is a journal that is freely accessible online and downloadable in a PDF format in several websites, such as https://www.ualg.pt/en/ualgoritmo-magazine (accessed on 15 December 2021), https://ualgoritmo.wixsite.com/website (accessed on 15 December 2021), amongst others. The UALGORITMO publishes concise communications in layman Portuguese language about the main results or activities previously described by researchers of the University of Algarve in mainstream indexed journals. Communications on any research topic are freely submitted via email by the authors to the editor of the UALGORITMO, who verifies that the main author of the work previously published in a mainstream journal is a researcher from the University of Algarve, and that the submitted communication only presents the results of the mainstream publication. Thus, the UALGORITMO contains review articles, formatted in a rather classical way, as many other magazines or journals with the same objective. Indeed, each article has a title, presents the authors and their affiliations, has an attractive graphical abstract designed to help the readers quickly gain an insight about the topic presented, a Portuguese abstract and its English version, an introduction, a section summarizing the results, and a conclusion to the communication. A reference to the original manuscript(s) serving as a base for the manuscript in the UALGORITMO, and previously evaluated and approved by expert researchers in peer-reviewed journals is also indicated.

As the communications published in the UALGORITMO are intended for lay public and not for expert scientists, the authors are instructed to submit a manuscript written in a very accessible language, using everyday terms. Noticeably, scientists are good at presenting their knowledge/work to their peers, but they fail or have more difficulties when it comes to communicating to a lay public [3,7]. Thus, to maximize the impact of this project, we implemented a revision process of the manuscripts submitted to the UALGORITMO by students (of the 10th to 12th year grade) from secondary high schools in the Algarve, under the coordination of at least one schoolteacher (Figure 1).

The role of the reviewers is to evaluate whether manuscripts under review are written clearly, and whether the message of the authors is understandable by them, who are a representative group of lay people. The reviewers are also asked to signal the sections in the text that are unclear and that should be revised by the authors. Additionally, reviewers may give suggestions to simplify the text and improve the communication impact of the manuscript by making it more accessible to non-scientific audiences. The reviewers may also suggest the addition in the article of the definitions of terms that are less common or unknown from the general public. The definitions appear in a glossary, on a left side margin in the final published version of the article, close to the place in the main text where the terms are used for the first time. Next, the report and suggestions of the high school reviewers are communicated to the authors to give them the possibility to make the necessary changes to the original text. The revised version by the authors is resubmitted to the high school reviewers for a final approval before publication. In a way, the high school students are introduced to a journal review process during the revision of the UALGORITMO manuscripts, and this activity has been perceived as a novel and enjoyable experience by them (Table 1 and Figure 2).

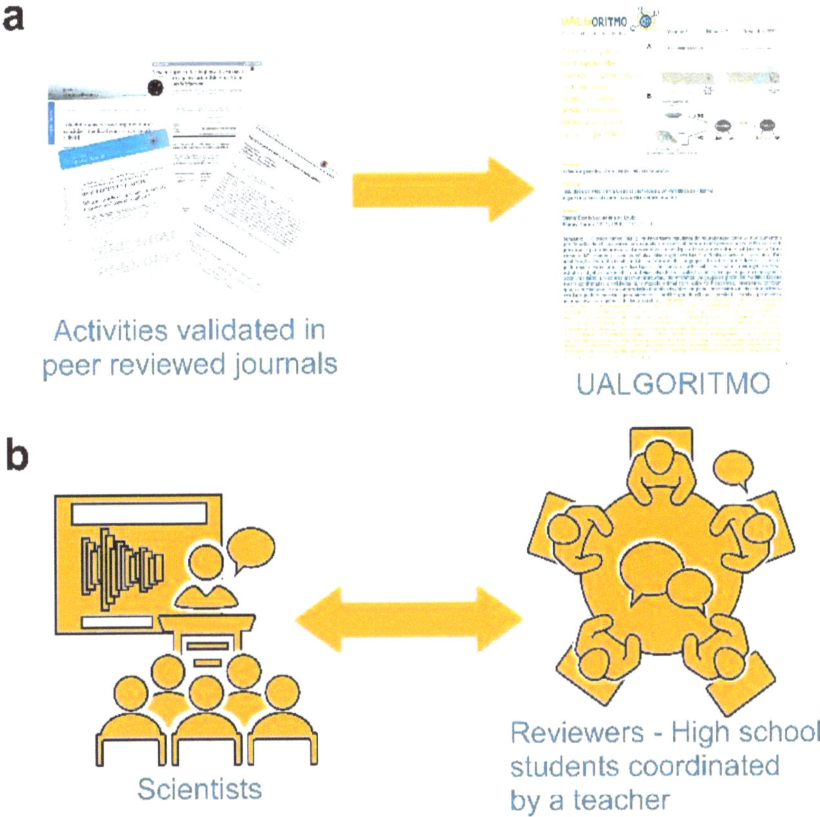

Figure 1. The UALGORITMO is an instrument for outreach activities to layman readers. (**a**) Articles published in peer reviewed journals (**left**) are the raw material for manuscripts submitted to the UALGORITMO (**right**); (**b**) The manuscripts from the authors (scientists), who often use a professional language difficult to understand by laymen, are reviewed for clarification of the text by secondary high school students under the coordination of a schoolteacher.

Table 1. Summary of the inquiry reports from high school students after the reviewing process [1].

Affirmations to be Graded from 1 to 5	Grade (%)				
	1	2	3	4	5
1—I enjoyed participating in the review process. 1 being equivalent to "not really", and 5 to "it was an excellent experience"	0	0	10	28	62
2—I found the process very difficult. 1 being equivalent to "not really", and 5 to "extremely complicated!"	31	41	24	3	0

Table 1. *Cont.*

Affirmations to be Graded from 1 to 5	Grade (%)				
	1	2	3	4	5
3—If there are more opportunities to participate in this type of reviews, I want to participate. 1 being equivalent to "not really", and 5 to "for sure"	0	10	7	28	55
4—The original text of the manuscript was clear and well written. 1 being equivalent to "not clear", and 5 to "very clear"	0	21	17	48	14
5—The scientific activity of the article is clearly exposed, and I understood everything. 1 being equivalent to "I didn't understand anything", and 5 to "very clear"	3	14	17	45	21
6—I liked the work presented. 1 being equivalent to "I didn't like it", and 5 to "I loved it, it was super cool".	0	3	14	38	45
7—I learned a lot from reading the communication I reviewed. 1 being equivalent to "I already knew all about that", and 5 to "I had no idea about these things"	0	10	31	34	24
8—In my opinion, the topic of the communication will interest many secondary high school students and many other people who are not scientists. 1 being equivalent to "I don't think so", and 5 to "yes, without a doubt"	0	7	21	48	24
9—After reading the communication I reviewed, I wanted to know more about the topic. 1 being "I wasn't very interested", and 5 being "I thought it was fantastic and I'm curious to see where this will end up in the future"	0	21	10	38	31
10—After reading the communication I reviewed, I felt like directing my future studies in that direction. 1 being equivalent to "no, it's not for me", and 5 to "studying and/or working in this area would be fantastic"	21	14	45	14	7
11—After reading the communication I reviewed, I wanted to know more about the research activities of the University of Algarve. 1 being equivalent to "no, it's not for me", and 5 to "yes I'm curious to know about the other things that are done at the University of Algarve"	3	3	28	41	24
12—After reading the communication that was revised, I wanted to know more about the educational training at the University of Algarve. 1 being equivalent to "no, I'm interested", and 5 to "yes I want to know more"	3	24	31	21	21

[1] The results presented are based on 29 inquiry reports. However, the inquiry reports are heterogenous, since some have been filled individually by each student involved in the reviewing process, while others represent the consensus of the whole reviewing group's opinion.

A small description and a picture of the reviewers are presented at the end of each article, as an acknowledgment for their valuable contribution to the revision of the manuscripts. The articles published in the UALGORITMO also contain a small biography of each author, and links to the websites of the research centers to which they belong, and links to the courses in which they teach at the University. To strengthen the outreach potential of the UALGORITMO, a website (https://ualgoritmo.wixsite.com/website, accessed on 15 December 2021) has been created to present all the articles published with the information about the authors and the reviewers, and direct website links to the secondary high schools which have participated in the revision and the Ciência Viva Center of the Algarve, which has largely contributed to the first contacts made with secondary high schools and for the launch of the UALGORITMO project.

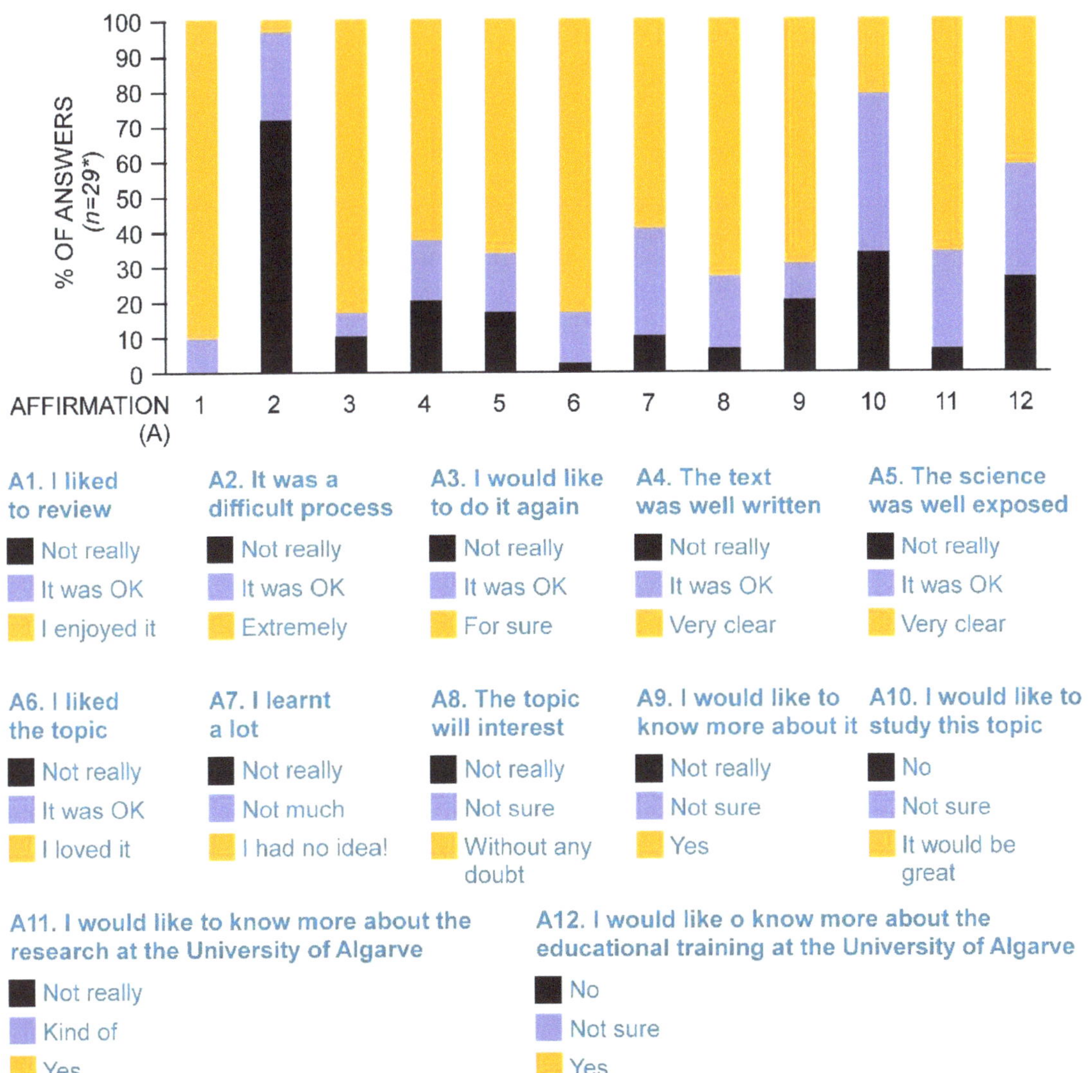

Figure 2. Graphical representation of the results from the inquiry after the review process. The results presented in Table 1 have been re-organized in such a way that answers 1 and 2 in one hand, and the answers 4 and 5 to in the other hand, which reflect nuances of the same answer were combined (*), are represented by black and orange histograms, respectively. The neutral answer 3 is indicated by blue histograms. The histograms represent the percentage (%) of answers. The summary of each affirmation and answers (A1 to A12) from the inquiry are presented below the graph.

3. The UALGORITMO, Designed to Maximize the Outreach

We believe that the reviewing process and the publication of the UALGORITMO maximizes the interaction between the University and all communities. Indeed, the first step of the outreach is the active and in-depth involvement of the high school students in the reviewing process, which implies understanding the work performed at the University of Algarve. The choice for the collaboration with secondary high school students, of the 10th to 12th year grade for the reviewing process was made because these students have sufficient scientific culture to understand the message of the manuscript and to provide a real input in the improvement and the clarification of the text for a larger audience. Of interest for the University, these students are only few years away, or months even, from leaving high school and make a choice for the next step towards their future education and professional activities. Surprisingly, many students who participated in the review process were unaware of the many courses and training provided by the University of Algarve, and they were also delighted to discover the research areas undertaken at the University during the revision process (Table 1 and Figure 2). Thus, the collaboration in the revision of the manuscripts for the UALGORITMO is an opportunity to introduce the educational structure and activities of the University of Algarve to these high school students, and perhaps provide them with paths that they did not envision before being in contact with the scientific activities and courses of the University of Algarve.

In a second phase, we believe that the students will be inclined to announce their contribution to the revision of the UALGORITMO manuscripts to family and friends, with a regain of interest in the divulgation, when the UALGORTMO is published online. This would propagate the outreach beyond the reviewers' circle, and perhaps motivate other readers to be in contact with the University's activities through the UALGORITMO. Although the UALGORITMO is freely accessible online by virtually anyone, printed versions of the journal are also sent to the teachers who participated in the reviewing process and to all libraries of the secondary high schools of the Algarve, which we believe will constitute another level of dissemination of the University's activities.

We have also encouraged the participation of Master's or Doctoral students from research groups, in writing the communications for the UALGORITMO, even if they were not authors of the original peer reviewed article. In addition to the authorship in the UALGORITMO publication, we expect that these University students will find in this process a way to train their writing skills, and to be incentivized for future scientific outreach actions. Moreover, the involvement of the research students will certainly alleviate the writing task of the senior author by reducing the time spent in the preparation of the manuscript, and this liberates Senior authors for the development of other activities. Of course, we also believe that these students will communicate their participation in the publication to friends and family, and even mention this activity in their curriculum vitae, which may be considered as a form of outreach, particularly if the reader of the curriculum will have the curiosity to discover the UALGORITMO. Additionally, the Professors at the University may also recommend the publications in the UALGORITMO to their pupils and junior students for an easy immersion in the research thematic. Of interest, the abstract written in English was introduced to give the opportunity to non-Portuguese readers to learn about the research topic addressed by the authors, and to find the original peer-reviewed article using the DOI presented in the reference section. This, in a way, provides another layer of dissemination of the original scientific report.

Finally, the launch of each new edition of the UALGORITMO is announced to the researchers and students at the University and to secondary high schools, and has taken place during meaningful events, such as the meeting of the Advisory Board for University of Algarve Training Offer, which has the strategic objective to bring the Algarve Academy closer to regional schools. This advisory board is constituted by representatives of the Councils of the Algarve, the General Direction of Schools and the Employment and the Vocational Training Institute. Thus, the launch of the UALGORITMO has also repercussions through the academic and regional communities present at this event.

Overall, the strategy surrounding the preparation, edition, launch, as well as the accessibility of the UALGORITMO contributes to maximize the impact of this journal to fulfil the scientific outreach expected.

4. The Reviewing Process of the Manuscripts and the High School Educational Mission

The rules for the revision of the UALGORITMO manuscripts were purposely left simple to provide a maximal autonomy to the reviewers. In essence the only rule is that the reviewers have to be students of the 10th, 11th, or 12th year grade and supervised by a teacher. Consequently, reviewing groups were composed, organized, and managed in very different ways according to schools and supervisors. Curiously, some reviewing groups were constituted by only one or two students, while others gathered a larger number of students, some of which were composed by up to three different classrooms engaged in different educational fields, such Biology/Geology and Portuguese, or Languages/Humanities and Biology/Geology. An interesting heterogeneous group was formed with students from different year grades (11th and 12th), as well as different educational fields and classrooms, such as Languages/Humanities, Sciences/Technology, and Socioeconomical Sciences. Some teachers have opted for the revision of manuscripts related to the fields that they teach, while others have not directed their choice to any particular field. This heterogeneity in the composition of the group of reviewers may explain the fact that some students are not inclined to continue their educational training in paths related to the topics that they have reviewed in the manuscript, although they enjoyed reading about the topic (Table 1 and Figure 2). Indeed, it is unlikely for a student engaged in Languages/Humanities to wish to continue in Sciences, for example.

Thus, multiple revision strategies were implemented by the reviewers, but we will briefly exemplify the modus operandi adopted by two groups. The first example comes from a reviewing group composed of a unique classroom, in which the teacher first instructed all students to the reviewing process. Next, each student received a copy of the manuscript for an individual reading to identify the central idea/message of the article. Then, all students together reviewed the article and discussed the concepts and definitions, explored the scientific question, highlighted the significant passages and information that complemented the central idea. In addition, the students collaborated to search for related articles and check whether these articles described data in agreement with the message presented in the manuscript under review. During the reviewing process, the students also learnt the bases for the construction of argumentative texts, and finally they reported to the authors the changes necessary in the text and whether relevant aspects were missing in the manuscript. The second example of the reviewing strategy involves two classrooms, Science/Technology and Languages/Humanities. Here, the revision process started with Science/Technology class, which was divided into several groups for independent analysis of the manuscript. After the initial analysis, each group presented a proposal for changes in the texts, definitions to be added to the glossary, and novel illustrations. All proposals from each group were then analyzed by the entire class to identify all unknown concepts and expressions. Next, a structured revision proposal combining all changes suggested, was communicated to the second group of reviewers composed of the Languages/Humanities students. This second group of reviewers proceeded the revision as described for the first group and sent their own comments back to the first group for a final agreement on changes to be proposed to the authors.

The program of the XXI Portuguese Constitutional Government in the law decree 55/2018 (https://dre.pt/dre/detalhe/decreto-lei/55-2018-115652962, accessed on 15 December 2021) and in the directives Profile of Students Leaving Mandatory Schooling (in Portuguese Perfil dos Alunos à Saída da Escolaridade Obrigatória, termed PASEO hereafter, approved by the dispatch 6478/2017, https://dre.pt/dre/detalhe/despacho/6478-201 7-107752620, accessed on 15 December 2021), has established the priority to implement student-centered educational policies in secondary high schools, by creating conditions for a balance between knowledge, understanding, creativity/originality, and a critical/creative attitude. The objective is for high schools to prepare students to develop skills that will allow them to understand and integrate new knowledge, to communicate efficiently and to solve complex problems that they may face due to globalization and accelerating technological developments. Undoubtedly, the diverse methodologies adopted by the reviewers for the revision of manuscripts to be published in UALGORITMO have allowed the students to develop practically all the skills listed in PASEO.

In addition, through its articles, the UALGORITMO may also be envisioned as an instrument to facilitate the very important and continuous actualization of the scientific knowledge and competencies that schoolteachers have to renew and acquire in order to efficiently transmit to their students the scientific literacy and skills needed to face ongoing and future societal challenges [9].

5. Discussion

The UALGORITMO is a free digital journal dedicated to the outreach of scientific, technologic, and artistic activities of the University of Algarve for the lay public. For a maximum impact of outreach by the UALGORITMO, the texts of the manuscripts submitted by the authors are tendentially written with simple day-to-day Portuguese language and revised by students at secondary high schools for further clarification. To simplify both the writing and the revision processes in the UALGORITMO project, we have opted for short manuscripts that summarize a main publication of the authors in a peer reviewed journal. Indeed, to summarize a published work is much easier and less time consuming for the authors than to write an original publication, and it is well reported that time is an issue for scientists to engage in outreach activities [8]. In addition, since the reviewers do not have to validate any science exposed in the manuscript, their task is easier, and they can concentrate their effort on the simplification of the text. In this respect, the project UALGORITMO is in line with other international science communication initiatives and outreach for society, such as "Atlas of Science" (https://atlasofscience.org/, accessed on 15 December 2021) and "Frontiers for Young Minds" (https://kids.frontiersin.org/, accessed on 15 December 2021), in which the articles published are written by scientists. However, in the "Atlas of Science" published layman's summary articles are not reviewed, and in "Frontiers for Young Minds" the manuscripts are revised by children and young people. The participation of the researchers of the University of Algarve and high school students to the preparation and revisions of the manuscripts to the UALGORITMO has been very enthusiastic and dynamic and has already allowed the publication of four volumes since its launch in October 2019. The UALGORITMO is freely available online to the general public on several websites either dedicated to the UALGORITMO project itself and at the University website, or on social networking sites for scientists, such as ResearchGate, dedicated to the community and researchers and students interested in science. Although the published articles have been read and downloaded thousands of times on the various websites where they are available, there is a lack of an instrument that can precisely determine their impact on the general public and students interested by science. In the future, to expand further the outreach, it would be interesting to develop complementary summaries of the work presented by the authors in short audio or video supports that would be available at the UALGORITMO website and other social media sharing platforms.

Apart from fulfilling the purpose of disseminating science and arts generated at the University of Algarve, the UALGORITMO contributes to the exposition and discovery of the University from a larger population. This is a very positive and important aspect since the University of Algarve was only founded a little over 40 years and has not yet reached the prestige of other Universities in Portugal, some of which have multiple centuries of existence. Furthermore, the University of Algarve is located in the far south of Portugal, far from the strong economic and cultural national cities that exert an attraction for many people, including students of secondary education of the Algarve who consider these places as their next path for future higher education. However, these students often disregard the University of Algarve because they are simply unaware of the activities developed at this University. The results obtained during this relatively short time of interaction with the secondary high schools for the purpose of revision of the manuscripts submitted to the UALGORITMO, clearly reveals this fact through the answers to the enquiries that the students have given. Interestingly, the results of these enquiries also indicate that a substantial proportion of these students have interest in discovering more about the University of Algarve. This is another very positive aspect of the UALGORITMO as a mean for outreach, developed in close collaboration with high school students, because it encourages a genuine curiosity for other activities developed at the University of Algarve, and it facilitates the access to websites containing all the information about research and teaching. In the future, it would be of interest to evaluate whether the UALGORITMO had an impact in the high school students' community for their ingression at the University of Algarve. Similar initiatives to the UALGORITMO could be adopted by other national or foreign universities that are outside the main educational scope to promote their visibility both at a local and national level.

Finally, we believe that the UALGORITMO also serves the high school education objectives delineated by the Portuguese educational system, since the reviewing processes of the manuscripts for the UALGORITMO involve methods and techniques that support students in building a genuine scientific, technical, and technological knowledge and culture, through the interpretation and communication of information and critical and creative thinking, which additionally contribute to values such as citizenship and engagement, curiosity, reflection, and innovation. Moreover, the UALGORITMO project is also very pertinent because students, acting as reviewers of articles, become more aware of how science is advanced and transmitted, and may be inspired to be involved in research in their future professional careers. On the other hand, they actively contribute to the dissemination science and its understanding by the public, by turning the message of a scientific article into an article that simplifies the scientific language, in a language more accessible to everyone, which allows for the promotion of scientific literacy in the community in general. This aspect is particularly important in Portugal, since studies carried out on the scientific literacy of the Portuguese population showed that decades ago the scientific knowledge of the Portuguese used to be lower than the average of the populations of other European countries. However, in recent years Portugal has been one of the few countries of the Organisation for Economic Co-operation and Development (OCDE) with clear improvements in youngsters' scientific literacy, which results from a Portuguese policy to improve teachers' qualification with a reinforcement of their scientific education programmes and implementation of lifelong training options for their career development and progression, and from an increasing effort of outreach engagement by scientific communities [10–13].

Author Contributions: Conceptualization, J.B.; writing—original draft preparation, J.B., S.F. and C.A.R.; writing—review and editing, J.B., S.F., C.A.R., F.d.R.C. and S.N.d.J.; project administration, J.B. All authors have read and agreed to the published version of the manuscript.

Funding: This work was supported by Fundação para a Ciência e Tecnologia (FCT) and the Comissão de Coordenação e Desenvolvimento Regional do Algarve (CCDR Algarve) for the project ALG-01-0145-FEDER-28044 "VITAL", PTDC/BTM-TEC/28044/2017.

Institutional Review Board Statement: Not applicable.

Informed Consent Statement: Not applicable.

Data Availability Statement: Not applicable.

Acknowledgments: This work is dedicated to the memory of Laura A. Bragança (12 May 2003–10 May 2019), who inspired the realization of the UALGORITMO, and to Mariana Isabel SLP Alves (9 July 1992–5 May 2008). We thank the authors of the articles published in the UALGORITMO, and the students of Secondary Education and their teachers for the reviewing process. Finally, we thank all members of the UALGORITMO Editorial Committee and Associate Editors, the staff of the Library of the University of Algarve with a special thanks to Salomé D'horta, and Cristina Viegas-Pires director of the Ciência Viva Center of the Algarve.

Conflicts of Interest: The authors declare no conflict of interest.

References

1. Entradas, M.; Bauer, M.W.; O'Muircheartaigh, C.; Marcinkowski, F.; Okamura, A.; Pellegrini, G.; Besley, J.; Massarani, L.; Russo, P.; Dudo, A.; et al. Public communication by research institutes compared across countries and sciences: Building capacity for engagement or competing for visibility? *PLoS ONE* **2020**, *15*, e0235191.
2. McClure, M.B.; Hall, K.C.; Brooks, E.F.; Allen, C.T.; Lyle, K.S. A pedagogical approach to science outreach. *PLoS Biol.* **2020**, *18*, e3000650. [CrossRef] [PubMed]
3. Brownell, S.E.; Price, J.V.; Steinman, L. Science Communication to the General Public: Why We Need to Teach Undergraduate and Graduate Students this Skill as Part of Their Formal Scientific Training. *J. Undergrad. Neurosci. Educ.* **2013**, *12*, E6–E10. [PubMed]
4. Wilson, M.J.; Ramey, T.L.; Donaldson, M.; Germain, R.R.; Perkin, E.K. Communicating science: Sending the right message to the right audience. *Facets* **2016**, *1*, 127–137. [CrossRef]
5. Kappel, K.; Holmen, S.J. Why Science Communication, and Does It Work? A Taxonomy of Science Communication Aims and a Survey of the Empirical Evidence. *Front. Commun.* **2019**, *4*, 4. [CrossRef]
6. Weigold, M. Communicating Science: A Review of the Literature. *Sci. Commun.* **2001**, *23*, 164–193. [CrossRef]
7. Besley, J.C.; Dudo, A.; Yuan, S.; Lawrence, F. Understanding Scientists' Willingness to Engage. *Sci. Commun.* **2018**, *40*, 559–590. [CrossRef]
8. Ecklund, E.H.; James, S.A.; Lincoln, A.E. How Academic Biologists and Physicists View Science Outreach. *PLoS ONE* **2012**, *7*, e36240.
9. Azevedo, M.-M.; Duarte, S. Continuous Enhancement of Science Teachers' Knowledge and Skills through Scientific Lecturing. *Front. Public Health* **2018**, *6*, 41. [CrossRef] [PubMed]
10. European Commission. European citizens' knowledge and attitudes towards science and technology. *Spec. Eurobarometer* **2021**, *516*, 1–4.
11. Cordeiro, A.M.R.; Alcoforado, L. Education and development. *Méditerranée* **2018**, *130*, 1–14. [CrossRef]
12. OECD. *OECD Skills Strategy Diagnostic Report: Portugal 2015*; OECD: Paris, France, 2015. [CrossRef]
13. OCDE. *PISA 2012 Results in Focus What 15-Year-Olds Know and What They Can Do with What They Know*; OCDE: Paris, France, 2012. Available online: https://www.oecd.org/pisa/keyfindings/pisa-2012-results.htm (accessed on 15 December 2021).

MDPI AG
Grosspeteranlage 5
4052 Basel
Switzerland
Tel.: +41 61 683 77 34

BioChem Editorial Office
E-mail: biochem@mdpi.com
www.mdpi.com/journal/biochem